Soul Mastery

*A Conscious Curriculum for Soul Evolution
& Energy Mastery*

*A 5th Dimensional
Quadrilogy of Wisdom on:*

**Energy, Vibration, & Frequency
*Conscious Neutrality
*Mastering Abundance
Soul Evolution

TONI G. BOEHM

First Edition 2019
All Rights Reserved Copyright 2019
By Toni G. Boehm, Ph.D.

Soul Mastery:
A Conscious Curriculum for Soul Evolution &
Energy Mastery

Inner Visioning Press
430 N. Winnebago Dr.
Greenwood, MO 64034
816-304-3044

The author would like to thank Mallory Herrmann
for her editorial expertise.
Published by Inner Visioning Press
Printed in the United States of America

ISBN # 978-1-64516-325-1
978-1-64516-325-1

1. Body, Mind & Spirit
2. Self-help
Title: Soul Mastery

Dedication

This book is dedicated to the evolution of consciousness, and to the idea that everyone can be a master facilitator of energy.

Forever and always, to my beloved Jay.

Acknowledgements

To Charles Fillmore, whose influence and mentoring changed my life. I never met you in the third dimension, yet your presence is tangible to me.

To Sophia and Paul, you introduced me to the idea of soul mastery as a "living" concept.

To all the teachers and masters who shared my soul journey, in whatever form you showed up, I thank you.

Table of Contents

1

A Shattering Experience

Imagine how the world will change when humanity realizes
its interconnectedness with all life and
works together to restore the...well being of the whole.[1]

~

It was a beautiful, bright, warm day in mid-October 2005. The smell of fall was wafting through the house. As I arose from my chair, having just completed my morning meditation, a mystical, transcendent, luminous, radiant, and life-changing event spontaneously occurred. In one singular moment, my life as I knew it, changed forever.

I realize now, after having traversed the spiritual path for more than forty years, that I have experienced hundreds of spiritual-awakening types of experiences. These include experiences such as spiritual quickening, mystical awareness moments, spiritual baptisms, ah-ha moments—when an interior lightbulb of awareness turns on, and so forth. This, event which I am about to share, was *NOT* one of those types of experiences! This was infinitely and completely different, and its effect remains with me today.

For several years, after the event, I did not speak of it to anyone, for I had no context or words to describe it; yet, its energy felt so real and familiar. I guess, deep down I thought that perhaps, people would think I was strange, or even crazed, if I described the event and its impact on my physical, mental, emotional, and spiritual states of being. For it did affect me on multiple levels of being, and still today, is energetic and vibrationally palpable.

The luminous event, as I lovingly refer to it, was an unexpected, unanticipated, and unabashedly a shocking experience that in an instant, dismantled and shattered elements of my small-self, my adverse ego, and my understanding of third dimensional awareness. In its transcendence, it opened the way for higher realms of dimensional awareness to awaken and to take-up residence within my soul and consciousness. The following is my self-proclaimed feeble attempt to portray this radiant, luminous, primordial, transcendent, and universal multi-dimensional-alignment event. Here is what occurred:

---

[1] Nassem Harramein, *Resonance Academy,* www.resonanceacademy.com/homepage

I was standing in my kitchen in full consciousness, taking a step forward, when in a stunning nano-second all third-dimensional awareness was replaced by infinite awareness. A radiant, luminous, stillness overcame my being. This radiant and luminous awareness arose, descended, encompassed the entire field called Toni. Within, without, around, above, and below, a shift occurred. It was sensory, perceptual, and cellular: every fiber of my being felt as if it were on fire and alive with a presence and energy, that has no words to describe "It".

Silence, stark stillness, and a spaciousness of emptiness enveloped the energetic field around the being known as Toni. Amid this field of pure, profound silence, an audible sound of glass shattering arose. I felt the protective shield of false beliefs and perceptions that had been held in place by third-dimensional constructs shatter and dissolve.

Then there was no me, there was no sense of a person named Toni, only a sense of no-thing, the primordial void, and a sense of a spaciousness of emptiness. I knew nothing and at the same time I knew everything. And I knew that I knew.

All was silent. All slowed to a space of pure, palpable stillness, with no-time, no-space, and everything radiated forth an intense aliveness, a luminosity, and a living spirit of light. A living energy was overwhelmingly present in the field as a palpable energy (I now refer to this energy as the *Universal Impulse*).

There was an awareness that Toni, the small personal, survival-oriented, egoic-self, had slipped aside or was gone. I do not know, for in that moment only, pure perfect profound peace existed. And along with that peace was a transcendent knowing of the interconnectedness of all things. All was ONE and was alive in a luminous unified field of consciousness. A harmonic resonance, a field of unconditioned love for all, was alive in the no-space.

In this field of pure unconditioned harmonic resonance, all persons who I had ever perceived as having wronged me paraded across an invisible screen and a knowing that there was nothing, no-thing, to forgive them for replaced all other thoughts and energies. Everything had dissolved, erased, lifted: only a pure harmonic resonance was alive in this luminous field of awareness. I was spoken to by an invisible voice, telepathically. Its vibration said, "This is divine resonance. This is what authentic love feels like." This love was a pure field or state of harmonic resonance, different from anything I had previously called, or known as, love.

In a nanosecond, this energetic field of harmonic resonance/primordial vibration/Universal Impulse took up residence as the primary impulse within my being. The Toni I knew, in that moment, was gone and all her wants and needs were dissolved. There were no needs, no desires, no lack, no-thing: only an awareness of a resonance field of pure being.

As suddenly as it began, it was over: was it a moment, a lifetime, 5 minutes, 20 minutes—who knows? But that palpable Universal Impulse remains with me. I now know, with a sense of clarity that feels profound and simple at the same time, that this Universal Impulse was always within me. It was just covered with layers of muck, karmic density, and other life-experience denseness, which, once sufficiently dissolved, allowed a great expansion of Its light and energy to surface and shine forth.

I now know beyond a shadow of a doubt that there is nothing to get; it is all here, present and alive, right now, in this moment. I now know I am now—in the midst of pure being and immersed in the living presence of light, life, and wisdom, I AM that I AM! And I know for you that all dimensions and levels of consciousness are available to you right here, right now! It can be no other way!

A few months after this experience I discovered the quote by Nicola Tesla; "*If you want to know how the universe works, think in terms of energy, vibration, and frequency.*" Discovering this statement serendipitously and dramatically shifted the trajectory of my life and shaped a new direction for my spiritual evolution and education, and soul expansion.

As I tapped into the hidden wisdom, life-shifting energies, conscious expanding, and palpable higher-dimensional vibrations concealed within the context of that simple phrase, my reality both expanded and shifted. My perspective on how to interact and participate with people, experiences, life, the universe, multidimensional awareness, and how to create a higher vibrational landscape of "living" reality, totally and completely transformed.

In many ways the entire experience was an unlearning process. I had to release all I thought I knew, in order to dance at the edge of mystery. I felt that everything that was unfolding was being guided by the newly grounded energy of the Universal Impulse; and that impulse, now felt like a dynamic, vibrating force of expansive energy, a "living" awareness.

Over time, my body system "recalibrated" and normalized into this new field of energy, it became a stabilized norm for me. With this new norm, I began to discover a profound sense of spiritual simplicity. It was from a paradox of this sense of spiritual simplicity, along with a sense of expanded awareness, that soul mastery, as a consciousness expanding practice and wisdom-teaching, was birthed.

I discovered that when the Universal Impulse courses through my being as a palpable energy or frequency wave, in these moments of movement, there is a dynamic, energetic "imprinting" going on in my energy field. The Universal Impulse is imprinting higher dimensional vibrations and frequencies of light, wisdom, knowledge, and awareness into, or onto, my field, consciousness, and/or cellular structure.

These imprinted vibrations and frequencies support my capacity to transmute, transform, and dissolve lower resistance-type energies—in alignment with my active and conscious claim and participation—for shift and change. My conscious alignment allows new energies to evolve and to bring new wisdom, ideas and information. That wisdom and awareness when unraveled often invites radical new insights and more expansive understandings regarding higher dimensional awareness and its impact on my life, soul evolution, and how-to be a conscious facilitator of energy.

*Energy and vibration in the form of thoughts, and feelings
are contagions—be careful of what you spread!*

~

There are a few ideas to take note of, as you engage with the consciousness-expanding information held within the pages of this book. First, throughout the book, a few ideas and concepts are offered more than once. This is intentional because repetition supports memory absorption and reinforcement!

Interlaced throughout the chapters, and as a final chapter, are practices entitled, *Alignment and Attunement Practices and Experiences (A&A)*. These *A&A* experiences are designed to be catalysts for: awakening; remembrance; expansion of awareness; bringing clarity, intentionality, and understanding to the subjects they support; acting as reminders to consciously and consistently practically apply the principles you are engaging with; and to vibrationally accelerate soul growth. Engage them knowing that you do not have to go anywhere special—to get spiritual—you only have to awaken to the splendor of the "living" Universal Presence and Impulse that is already within you!

Woven throughout the pages is my story of conscious evolution. It is my story; your story may look different. My path is a path of mysticism and mystical awakenings, therefore, I grow and evolve through those types of experiences. You will always experience what you need, in the way you need it, that is how the process works—trust it!

Know that all levels of awareness, levels of dimensions, all light, all initiations, and all information required for soul mastery, spiritual growth, and soul evolution are available to you, right here, right now, in this present moment. It can be no other way! Make your choice to engage with the stimulus in a manner that promotes soul growth, and not the continuation of status quo!

This book is an active transmission of energy and inherently holds a quickening vibration. As you consciously choose to engage in the discovery and realization of its message, you will be spiritually quickened and alchemically transformed (like ice transforms into water under the right conditions) and, ultimately, aligned with new levels of vibrational and dimensional awareness. Evolution awaits!

A&A: Universal Alignment Invocation

<u>Instructions</u>:
1. Repeat a minimum of twice a day, once in the morning upon arising, and then again, before you go to bed. Do this for a period of no less than 21 days.
2. Please note that the word <u>person</u> is not listed in the first paragraph of the *Invocation*, this is deliberate. Do not add it!

From a higher dimensional awareness, and metaphysical teaching perspective, you never request that a person be released from your life. Instead, you ask that you learn the lesson that the person is there to teach you, and to learn it quickly.

For once the lesson is learned the person will leave, or you will be so transformed that their presence doesn't bother you any longer.

Universal Alignment Invocation

The swift powerful activity of Universal Intelligence
now releases from me
all thoughts, beliefs, behaviors, ideas, and energies
that <u>are not</u> in vibrational alignment with
the universal principles of truth.

The swift powerful activity of Universal Intelligence
now attracts to me
all persons, thoughts, beliefs, behaviors, ideas, and energies
that <u>are</u> in vibrational alignment with the
universal principles of truth and the
highest possibilities in the unified field of consciousness.

2
Soul Mastery: What Is It?

*Spiritual evolution occurs as the result of removing obstacles
and not actually acquiring anything new.*[2]

~

After the luminous event, the primordial, transcendent, mystical, and fully lucid vibrational event, three things immediately occurred. First, an energetic doorway to a higher dimension of awareness opened within my being; I just knew certain things and I knew that I knew. Second, as mentioned, a universal impulse, a vibrational presence, took up residence within my being. This universal impulse was and is a palpable stream of luminous energy, that is steady and mighty.

It remains, even today, <u>as</u> I remain in alignment and non-resistance. I share this last caveat, because in the name of transparency and full-disclosure, there are moments in which I move out of the energy of conscious alignment, and into resistance and negativity. However, the recognition of the energy of resistance and negativity is now almost instantaneous, and so is my conscious decision to choose to realign.

How I showed up to life after the luminous event, shifted. The types of choices I made changed, and along with that, the landscape of my reality transformed, as well as how I viewed my life experiences. I began to realize that all the experiences of my life were not happenstance. There was something bigger going on.

Third, synchro-divinity happened! I began to observe that experiences, unforeseen, unanticipated, unplanned, astonishingly extraordinary or stunningly simple, and yet, all totally unexpected, began occurring. They came in the form of unsolicited chance meetings, serendipitous moments, out-of-the-blue intuitive hits, stunningly life-shifting meetings, and other catching-me-by-surprise moments, and they were happening to me, almost daily. It was if a universal energy or force was aligning experiences, for my highest spiritual good.

In honor of the spiritual impact each moment had on me, I named these unbeckoned, unsummoned, synchronistic, and grand-possibility attraction moments; synchro-divinity!

---

[2] David Hawkins, *The Eye of the I*, Veritas Publishing, Sedona, AZ.

Knock, And He'll open the door
Vanish, And He'll make you shine like the sun
Fall, And He'll raise you to the heavens
Become nothing, And He'll turn you into everything.
Rumi

~

One such moment of synchro-divinity that occurred shortly after the luminous event, came in the form of an unexpected phone call. A man, who I did not know very well, called and invited me to spend a few hours in the Archives at Unity Village. He had gotten permission to peruse some of the unpublished Fillmore manuscripts, and because of my deep heart-felt respect and love for Mr. Fillmore, he thought I might enjoy the experience. I joyously accepted his offer.

While there I discovered an unpublished 600-page, manuscript by 20th century mystic, Charles Fillmore. The moment I touched the pages of the manuscript it began to vibrate in my hands. In that moment, I knew that part of the next step on my soul journey was to prepare this manuscript for publication. To share the contents of it with those who were ready for its message. I knew it and I did it, within six months the manuscript was published!

Another energy-moving, spiritual, synchro-divinity experience was a conscious-lucid-dream interaction with the master teacher Jesus, and Charles Fillmore. This unraveled multi-dimensional levels of awareness within me and created an indescribable level of knowing and trust for the "awakening" process. Then there was a profoundly formative synchro-divine experience, which was the discovery of a quote by Nicola Tesla; "*If you want to know how the universe works, think in terms of energy, vibration, and frequency.*" This quote, and the subsequent unraveling of what it meant metaphysically and energetically, beyond the words written on the paper, was consciousness-altering.

Not so long after the Tesla quote unraveling, there was a moment of synchro-divinity, that was an audible call, the universal impulse-Spirit nudged me to reread the *Sermon on the Mount*! Saying yes, to that call, rocked my world, as it was a multi-dimensional spiritual experience (this experience will be explored in depth, in another chapter)! Also, there was a dream that revealed the name of a fourth-dimensional teaching and spiritual practice—Conscious Neutrality.

The last synchro-divinity experience to be shared, was an unexpected moment of clarity and realization! In a lucid day-dreaming-like state, a vision spontaneously arose. I saw hundreds, upon hundreds, of dangling, multi-colored threads. Each thread held an idea, or a wisdom-teaching, that was the end-result of a lesson that I had learned; over eons of time. Suddenly, the threads began to pulse and expand. As the threads were expanding, they were also energetically, being lifted across dimensions, 3rd—>4th—>5th, and higher. Effortlessly, the threads moved across timelines, frequency ranges, and dimensions; their colors getting more luminous and brilliant, as they rose.

Instantaneously, the upward trajectory appeared to slow, and the threads began to invisibly coalesce and weave themselves into, what appeared to be, a multi-textured, multi-dimensional tapestry of luminous, golden and white light. That tapestry, an invisible, energetic, light-filled, soul-memory-picture became a touch-stone for the creation of the quadrilogy of wisdom teachings—quadrilogy, is a group of four—known as, Soul Mastery.

The dazzling effects of seeing lights and visions
are not necessarily signs of spiritual growth.
They are only external indicators of movement of energy.
Jack Kornfield

~

The quadrilogy that forms the Soul Mastery collection includes:

1. Energy, Vibration, & Frequency[3]
2. Conscious Neutrality
3. Soul Evolution[4]
4. Mastering Abundance[5]

Each one of the quadrilogy's individual wisdom-teachings, is a guidebook that holds a stand-alone set of instructions and spiritual practices that creates a unique perspective on how-to unfold the soul's innate wisdom for evolution and energy mastery; in accord with the specific foundational premises and instructional design processes of their subject matter.

As you peruse the quadrilogy, allow time for the energy of each wisdom-teaching to absorb, assimilate, and evolve its unique ideas in you; for these wisdom-teachings are consciously designed and arranged to support an expansion of awareness! Additionally, in support of absorption and clarity, a few of the ideas and concepts contained within

---

[3] Toni G. Boehm, *Mastering Abundance*, Inner Visioning Press. Amazon.com
[4] Toni G. Boehm, *Conscious Living*, Inner Visioning Press. Amazon.com
[5] IBID

the quadrilogy, are offered more than once. Repetition supports absorption and assimilation!

You cannot be a Master and a victim, at the same time!
Paul Selig

~

Soul mastery, as a quadrilogy of leading-edge wisdom-teachings, seamlessly weaves both quantum science and spiritual principles into a conscious, organized, progressive, systematic, and transformational soul-expanding experience.

As such, the quadrilogy— is an intentional, experiential, evolutionary, consciousness-expanding, awareness-elevating, energy mastering, quantum-oriented, unlearning, alignment, attunement, assimilation, absorption, integration, soul-awakening, and higher-dimensional experience that—vibrationally realigns all aspects of being—physical, mental, emotional, and spiritual—from the inside out—not from the outside-in.

Soul mastery is a shift out of a third-dimensional human-ego-survival-dominated orientation and reality into a multidimensional, universal, and quantum-oriented existence. However, it only works as the teachings are consciously, and practically, applied.

"In a seven-tone scale the eighth note is the octave,
twice the pitch of the first note,
and so, signals the movement to a new level.
This may be why, in religious symbolism,
the eighth step is often associated with spiritual evolution...[6]

~

As the soul mastery quadrilogy supports you in aligning with higher dimensional awareness (4th, 5th, and higher) to understand the idea of dimensions and higher dimensional awareness; imagine a piano. The piano holds a range of notes that exist in different octaves, lower octaves and higher octaves. It is the octave you choose to play in that affects the outcome of what the music will sound like.

The octave chosen creates the predominant field of harmonic resonance. Lower octaves create heavy, dense, and base-like sounds, whereas higher octaves create lighter sounds. No octave is right or wrong, it all depends upon the type of music you desire to play.

---

[6] John Martineau, *Quadrivium: The Four Classical Liberal Arts of Number, Geometry, Music, & Cosmology*

It is the same with you, the dimension (octave) you choose to "play" in, becomes the dominant resonance or vibration that affects the manifestations that are created in the landscape of your reality. Choosing to play in the third- dimensional octave orients you towards a reality laden with separation, survival, fear, anxiety, greed, and being unconscious of who you really are as a spiritual-vibrational-energetic being.

Choosing to play in consciously in a fourth-dimensional octave, holds a massive leap in consciousness. Tapping fourth-dimensional awareness initiates: a movement of energy between dimensions, for it is an energetic bridge between them; a more metaphysical understanding of life; alignment with conscious transformation and transmutation of dense and shadow energies; alchemical transformation through the conscious demonstration of spiritual principles; conscious energy facilitation and mastery; synchro-divinity abounding; conscious neutrality; an understanding of the role of energy, vibration, and frequency in soul evolution; and more.

Choosing to play in a fifth- dimensional octave orients you towards a reality that holds: Unity-Christ Consciousness; instantaneous manifestation; oneness; wholeness and life; telepathy; intuitive knowing; clairvoyance; mastery at all levels; a body filled energetically with light, for it has no fear, anxiety, or dense energy trapped within it; and much more.

The dimension or octave you occupy determines the experience you have, your reality reflects your level of consciousness. Wherever you go, you take your consciousness. You get to choose whether you play in an environment that is harmonious, vibrationally aligned with the unified field of consciousness, or one that is more discordant.

> *The object … is to discern truth and prove it.*
> *Jesus and his followers did not work miracles,*
> *they applied laws in a field of mind*
> *beyond that of average men.*
> *These laws constitute what is called*
> *esoteric Christianity, [4th Dimension], or mysticism.*
> *The time has now come for all…to come up higher.* [7]

> ~

Soul mastery simply defined, and with each word reflected upon separately, starts with soul. Soul is the sum total of all awareness—conscious and unconscious. Soul includes, metaphysically, the three

---

[7] Charles Fillmore, *Unity Magazine*, February 15, 1919

phases of mind[8]: superconscious mind, subconscious mind, and conscious mind.

Mastery, or a master, is referred to in this context as one who has mastered a skill-set or teaching. It is said that it takes about 10,000 hours of study and apprenticeship to master a skill or teaching.

So, a spiritual master, would be one who has developed the spiritual mind discipline, the capacity to hold steadily and with steadfastness the energy of higher dimensional awareness, even during the most challenging appearances and experiences.

Additionally, in the midst of challenging experiences they have the capacity to be an agent of alchemical transformation of energy in a space or room and to call forth elevated emotions such as, trust, peace, joy, appreciation, etc.

If you do not transform your pain, you will transmit it.
Richard Rohr

~

Soul mastery as a spiritual developmental process will initiate a transformation in your awareness regarding a conscious claim and understanding of your personal relationship to the sacred source energy that courses through out your being.

If the word God, Spirit, or... makes you uncomfortable, choose a word that matches your current level of comfortability. There are a multitude of ways to refer to the source energy that lives and breathes you. You may choose to call it; Universal Presence, It, God, Source, Spirit, One, Oneness, Christ, Energy, Universal Impulse, Christ Consciousness, Christ Nature, Unity Consciousness, 5th dimension, Holy Spirit, Higher Dimensional Awareness, Path of Ascension, Self, No-Thing, I AM, Spirit, Buddha Nature, Source Energy, etc.

Why is it important to have a word you can say and be comfortable with? Because words have energy and power! And if you are shunning or reacting to a name, or still holding anger that arises from your religious past, you are holding onto the dense energy of old memories. And that density anchors you into 3rd dimensional energy—no matter how much you think you have spiritually grown.

What you hold on to, holds you back! What you resist, persists! Density weakens, not strengthens! Energy not cleared, is energy feared!

Ultimately, what you call this source energy, does not matter. What does matter is the energy and vibration that you feel, sense, and

---

[8] Ed Rabel, *Metaphysics 1*. Amazon.com

align with when you are speaking or thinking about source energy. Energy doesn't lie—energy not cleared—holds you hostage to old programming and resistance!

Part of the transformational process of soul mastery, is the opportunity to recognize that everything is energy, vibration, and frequency, at play. In discovering this, you will find that your relationship with this sacred source energy is first, and foremost, a vibrational relationship. You will realize that you are a vibrational-spiritual being having a human experience.

All persons in rare moments catch glimpses of the creative plan
as a whole, and of man's importance in its beauty and perfection.
But this subject is so deep and so far-reaching that
it can only be realized in small degree and
only by those who have developed spiritual sight and feeling,
and practice thinking in the fourth dimension.[9]

~

Soul is *the sum total of all awareness*—conscious and unconscious—and it includes what Mr. Fillmore called, the three phases of mind[10]:
- Superconscious Mind (SCM)
- Subconscious Mind (SubC)
- Conscious Mind (CM)

Each phase of mind is composed of various levels of energy, vibration, and frequency, and each phase of mind vibrates at different levels of dimensional awareness in accordance with the information, light, energetic density, and frequencies it holds; e.g., Superconscious mind holds 5th dimension and higher frequencies, Subconscious mind is pivotal, and works with the dimensional energy you program into it.

Thus, each phase of mind has a specific role in the unfolding of the experiences that determine your soul evolution, your soul story, and, ultimately, soul mastery. Therefore, it is imperative to be conscious, aware, and to observe, to the best of your ability, which mind you are engaging when you are making choices.

"If you obsess over whether you are making the right decision,
you are basically assuming that the universe will reward you for one
thing and punish you for another. The universe has no fixed agenda.
Once you make any decision, it works around that decision. [11]

---

[9] Charles Filmore, *Atom Smashing Power*, pg. 62
[10] Ed Rabel, *Metaphysics 1. Original copy, re-published by Toni G. Boehm, Amazon.com
[11] Deepak Chopra, *The Book of Secrets: Unlocking the Hidden Dimensions of Your Life*

The Three Phases of Mind

Superconscious Mind
*Christ—I AM
*5th D. & Higher
*A vibrational reality

Subconscious Mind
*SCM & CM imprint on SubC
*Individual & collective
story imprints here
*Connected to 3rd & 5th D.
*Is pivotal: can take
direction from 3rd, 4th, 5th
*Computer-like
*Yes, I have had this
experience before!

Conscious Mind
*Attached to 3rd D. & Linear-
Dense Reality
*Judges by Appearance
*Freewill to Choose Change

The superconscious mind vibrates at a high-dimensional frequency as it is your connection with Source, Spirit, I AM, Universal Energy, etc. It is a constant and continuous connection to higher realms and dimensions of awareness that lives within you, whether you are consciously aware of it or not. Additionally, regardless of what you are feeling or experiencing in the moment, your connection to superconscious mind—I AM, Christ, Universal Presence, Spirit—is never (and will never be) broken. The superconscious mind is eternal and unchangeable, and it holds information as to all that you are from across time and dimensions.

From the beginning, humankind was designed to be in conscious connection with the superconscious mind and it was to be the creator of life experiences—the proverbial Garden of Eden experience. However, due to our free will, other choices were made by the collective and this allowed the adverse ego to set up shop through the subconscious and conscious mind relationship. Now, the true essence of the spiritual journey is one of consciously returning or reengaging with that pristine state of superconsciousness-I AM energy, as the dominant vibration for guidance. Right now, it is the conscious mind connected to the subconscious, that is dominant, with only an occasional input from superconscious mind.

The superconscious mind holds the Truth of who you are. However, due to millennia of forgetting and dropping into separation consciousness, you must make a conscious choice to return to the superconscious mind being the master of your energy field—not your emotions or past experiences.

From this superconscious level of awareness—which is universal, interconnected to all, connected to the unified field of consciousness, multidimensional, and spans across all time—you can access all memory, knowledge, wisdom, information, gifts of Spirit, and the attributes of God. These gifts include telepathy, extra-sensory perception, intuition, instantaneous wholeness, the cloak of invisibility, overcoming the elements, instantaneous alchemical transformation of energy, and much more.

Hidden, as a mystical teaching in the *Sermon on the Mount,* the master teacher, Jesus reveals how to align and attune with the super-consciousness mind, higher dimensional awareness.

We cannot selectively numb emotions, when we numb the painful emotions, we also numb the positive emotions. [12]

~

The subconscious mind is the great repository of your experiences, it is a data collector. From the age of newborn to seven, your subconscious mind was conditioned with programs by your parents, grandparents, culture, society, and more. It is these programs that you create from today and that still impact your behaviors.

The conscious mind is creative, and brain research shows it works creatively with you about 5% of your of day. But what happens the other 95% percent of the time, where does the mind create from? It creates from the data-programs available in the subconscious mind. If this is so, then what are you defaulting to for creativity? Old programs and habitual conditioned responses. This will continue until make a conscious choice to change your conditioned programs.

Charles Fillmore says the subconscious mind is like a old-time, photographic sensitive plate; it only imprints what you place onto it. And you placed a lot of conditioned data in it, in your early, formative years; when you were being taught who you were/are. These thoughts and ideas were not yours, but someone else's ideas and ideals of how the world works, and how you should respond to it.

The subconscious mind lies behind the superconscious mind and the conscious mind and is pivotal, it has the capacity to imprint

---

[12] Brene' Brown, "*The Gifts of Imperfection: Let Go of What You Think You're Supposed to Be and Embrace Who You Are*".

information from either the superconscious mind or the conscious mind, and to take direction from either. The subconscious mind is a repository, and a storehouse of data and your individual experiences, along with the experiences of the greater collective. Additionally, it does not know right from wrong—it only knows yes this happened, no it didn't; yes this is stored here, no it isn't. Thus, it is like a computer: stuff in, stuff out; garbage in, garbage out!

The subconscious mind holds both individual and collective information and energy. Over time, you create collections of similar type of energetic experiences and as they gather and hold enough energy, you create beliefs and perceptions out of them. Once you create a belief or perception, human nature kicks in with its need to be right, and you give your belief great meaning. You do this in order that, when the experience arises the next time, you can say, see I told you this always happens to me. It is these embedded neural pathways of meaning that run your life as unconscious, habitual behaviors, instead of Spirit. The subconscious mind is not logical, but it is literal, and it only knows yes. Yes, that memory or belief lives here, or it does not. It is more powerful than the conscious mind as it directs multiples of activities simultaneously and while doing this, it often confuses similar with the same.

Meaning, if something is happening to you, like meeting a person and not feeling safe, the not feeling safe is happening because the subconscious has sent a warning ("This same feeling or thing happened to you twenty years ago—beware!") when it is not the same thing at all. It may be similar in context or how it looks, but it is not the same. In the end, your reaction to the similar situation will usually be the same as the one you had twenty years ago, so it takes things very literally.

Often the subconscious mind—and this is important—confuses what is imagined with what is real. You imagine that someone doesn't like you and the subconscious mind begins to make it a real thing in your mind. Therefore, it automatically and impersonally works to achieve what is programmed into it. Success or failure, it doesn't care—you have programmed it and it does its job.

The subconscious doesn't like change—neither does the ego. It prefers status quo. Ever try to diet? Sticks for a day or two, then you capitulate back into known old unproductive ways. Yes, the subconscious is at your service.

The subconscious mind only supports yes! It ignores no! Because the subconscious doesn't know no, it drops no. If you say over and over,

I don't want to, the subconscious hears, I want to! So, it goes in search of what you want. It is yes-oriented. And when confronted with life experiences, the subconscious mind left to its own devices, goes in search of all the information you have stored, to substantiate, and confirm, a yes; even if it a negative experience.

Yes, I know this—this is familiar—and I know how to react to this—since I have experienced it before! It then sets off all types of neurochemical reactions to stop you from going forward, or to sabotage your goals, if you allow it. The subconscious is, in actuality, your minds default mechanism, for it allows you to go automatically on autopilot. In a nano-second you are back, entrenched in old programs and ways of being, without thinking!

The subconscious enjoys the status quo, and yet it is the home base for dreams, and it is the director of emotions and emotional responses through the controlling of physiological mechanisms, such as emotional response. Emotions were designed to be indicators and calibrators for you to weigh in, honestly, on where you in any given moment, or situation. They were meant to support you in recognizing the need to recalibrate yourself when things appear to be getting hairy or uncomfortable. However, instead of emotions' original design as indicators to implement recalibration, emotions are now utilized in the creation of meaning for your experiences. This undue influence has allowed your emotions to become the ruler of your being instead of intuitive guidance, or Spirit.

Remember the example of not feeling safe in a person's presence? Instead of being aware that a sense of a feeling of unsafe is arising, you make it mean something. "Oh, I don't feel safe. This person is dangerous. What are they going to do me? I remember another situation like this, and it is not safe. S/he is not safe!" Then, in order to be right, you go in search of evidence and/or people to collude with you to assure your meaning-making is correct. Meaning-making, is done in order that you might feel like you are right and to feel justified in your rightness! This is not about right or wrong, it is an invitation to notice and observe when you are participating in this type of behavior.

The subconscious mind has more childlike ways than adult-oriented ways. It often lives through the mind of the five-year-old in you. Anything that has traumatized or negatively impacted you from the past often becomes the basis of misperceptions that, in turn, create negative behavioral patterns that you unconsciously live out.

Vulnerability is... having the courage to show up and be seen when you have no control over the outcome." [13]

~

The conscious mind is important for several reasons. First, it is the point of freewill, it is where you make choices (conscious or unconscious) and decisions through judging by what is happening in the moment, etc. Second, it connects you to all the appearances of third-dimensional reality through your senses. It takes those sense perceptions and makes assessments in conjunction with the subconscious mind and makes decisions for action based upon those assessments. And usually, its decisions for action are usually reactionary—at least until you start to practically apply spiritual mind discipline.

Conscious mind manages your freewill. You were meant to take direction from the superconscious mind, but over time, through your capacity for freewill, you choose to "see" and "judge" life through the conscious mind, and the world of appearance. Freewill, unconsciously uses what the the conscious mind sees as the barometer for your life experiences, and as the dominant source of consideration for your thoughts, feelings, and emotional responses. Conscious mind also tends to support a critical and judging mind, which is aligned with the 3rd dimensional adverse ego, which likes to be in control, and is willing to manipulate circumstances and people. This is done for many reasons, the biggest of which are to feel safe, to feel liked, to be right, and to not be abandoned. Conscious mind, however, determines choice and can be trained to "see" things differently.

The conscious mind can multi-task, but its attention span is often limited, unless it is trained to be intentional and aware. It can and does use logic, analytics, and rationality, but through the connection with the subconscious mind these are usually steeped in the past and thus you make your decisions based on unconscious past perceptions. Thus, conscious mind is a reflection of your conditioned, subconscious mind. Your subconscious mind is conditioned, or indoctrinated with beliefs, by the age of seven. This conditioning was created by your parents, society, culture, heritage, tradition, nationality, teachers, past experiences, and more.

This conditioning is reflected by the choices your conscious mind makes as an adult, until you consciously choose to retrain it. Which

---

[13] Brene' Brown, *The Power of Vulnerability*, TED Talk

takes conscious work! Science is now revealing that you only intentionally use your conscious mind, for creativity and conscious choice, about 5% of the day, the rest of the time you default to the subconscious mind and its old habitual programming.

For instance, when you wake up each morning, you engage the conscious mind for a moment. Then you allow it to default to the recycling of the memories and habits of the subconscious mind. In fact, it is said that 70-80% of your day is spent in thought that recycles memories from yesterday. For instance, you get up on the same side of the bed, immediately recall how angry you are at so-and-so, then you review how much you hurt over what so and so said, then drive the same roads, stop at the same place for coffee/tea/eats, and on and on it goes.

Memories and habits are a record of the past and they have an emotional record attached to them. The stronger the emotional reaction of an experience, the more record and recall you have of it and the stronger is the emotional residue that gets embedded in your body.

Over time this emotional program becomes hardwired in your brain's circuits, and since you can't think greater than you feel, so these memories create a state of being in you—you be — sad, anger, the widow, the one picked on, not good enough, etc. And the familiar past creates a predictable future. It is said that engaging in routine habitual behaviors over and over actually reduces the freewill of the conscious mind. Is that what you really want for your life?

As one heals, we all heal.

~

When you awake and become the observer of self, you begin to awaken from the dream of the subconscious, the sense of separation, and the perceptions from the past; all of which have held you hostage. You begin to make new choice. You start the journey to being a master of energy facilitation.

As a master facilitator of energy, you notice the emotion as it is arising, and you are more interested in it as an observer and indicator of what is <u>going on in you</u> in the context of this experience. Thus, you use the emotion as a tool of observation and discovery. Then you choose to observe your reaction and discover what in you is at the root of it, rather than projecting the negative reaction or emotion onto someone else as the cause.

Involution (in-folding/receiving a thought)
always precedes evolution (unfolding).

That which is involved in mind evolves through matter." [14]

~

As a conscious growth process, soul mastery involves both involution and evolution. Prior to anything being manifested, there must first occur in mind an involution or downloading of an idea, question, revelation, reflection on a behavior, understanding, potential possibility, etc.

If you are observant, you will notice that involution is happening all the time. If you stop and take notice for a brief period, as little as 24 hours, you will notice how often idea downloads—involution—occur in the course of a day. Whether it is an idea about starting a business, going back to school, joining a group, stopping a behavior, choosing another route, making another choice, treating someone kinder, or doing something creative ideas are always flowing to you. However, the big question is: What do you do with them?

My personal journey has been filled with great ideas and incredible revelations. What I have learned is that if an idea download comes and I sense that it feels right, I tend to engage it, or do something about putting it into action, no questions asked. I take some forward movement on the idea. Whether the idea is big, like writing a book or starting a publishing company, or small, like calling a friend or writing an email, I will take some small action to move the idea forward.

Over time, by consciously saying yes, and being willing and receptive vehicle to receive new idea downloads, I now receive intuitive-type mystical messages on a regular basis. Those messages and ideas show up in my dreams, and support me in soul unfoldment or the development of new projects. Another thing that happens in the dream realm, is that I often meet wisdom-givers. The information that these wisdom-keepers share with me in this dream state is usually profound.

On several occasions, I have discovered that these wisdom-givers are actually famous teachers from other traditions; some living, some not, some I have been familiar with, and some not. These life-changing dreams have involved people such as, Swami Muktananda, Jesus, Charles Fillmore, Joel Goldsmith, Gurumaya, the Black Madonna, and more. All of these types of dreams, however, have held one thing in common: the involution of an idea. An idea that was a catalyst for a new action, a new direction, or a new directive for my journey—often wrapped in a metaphor, or a mystical saying I had to unravel. Following

---

[14] Charles Fillmore, *Revealing Word*, Involution, Unity Publishing

through on the message, always evolved me in ways that were beyond my wildest imagination.

For instance, the dream encounter with Jesus and Charles Fillmore led me to read the *Nag Hammadi Library;* and that sparked a desire to look for mystical teachings that might be hidden in context of the *Christian Gospels.* Each acted as a part of a progressive unfoldment process, revealing a whole new level of knowing and understanding regarding my soul's journey.

Engaging with the *Nag Hammadi Library,* along with the manuscript by Charles Fillmore, also awoke a deep remembering of hidden wisdom that was already alive within the depths of my being. Each, in its own way, was a catalyst for sparking a greater understanding of the role that the master teacher Jesus played in the cosmic evolution of humankind's soul growth.

The template for soul maturation and mastery (a.k.a., Christ template, fifth-dimensional blueprint) hidden in the story of Jesus is nothing less than a masterpiece, a complete soul evolution blueprint and roadmap to soul mastery (refer to the chapter on Soul Evolution). I would not have <u>seen</u> any of this without having first, raised my own vibrational awareness.

It would be a fruitless task to count how many times I had read the words of Jesus <u>without</u> seeing the mystical message hidden beneath the words. Then one day, the words came alive in this new context. How, did this happen? Why, did this occur? I had changed vibrationally and shifted into a new frequency of awareness, and through this new lens of awareness I now saw them differently. From a higher vibrational frequency and awareness, their hidden mysticism began to unfold before me. Remember, the level at which you choose to engage is up to you, along with the amount of time, energy, effort, and money you are willing to invest in yourself as a spiritual vibrational being on a journey to Self-realization. And your life experiences will reflect the "fruit" of your investment and effort!

> *God said ... Let us make them in our image [I AM] ...*
> *God blessed them... and gave them authority and*
> *dominion over all things, and the power to subdue...*
> Genesis 1: 26 – 28

~

Are you aware that the compelling drive, the primary force of the Universe is to evolve, is evolution? Do you realize that you are an evolutionary response to a need of the Universe? The Universal Impulse and Intelligence, in order to evolve a certain required aspect of the evolutionary process, created you! Therefore, you are Universal Impulse's answer to an evolutionary need!

What is that need? It is up to you to awaken and discover the true answer to that question, for yourself. Certainly, traversing the spiritual path to soul mastery supports you in this discovery process as the journey of soul mastery is one of continual discovery; remembering; awakening; claiming of who you are in dominion as the I AM Self; and clearing and transforming through the dissolving, overcoming, and releasing of embedded, individual and collective, beliefs and perceptions held in place by dense, old energies; third-dimensional fear and survival-oriented energies.

Soul mastery starts out as an appearance of you evolving self for self, but in the end the result is that you are evolving the individual self, known as you, in order that collective consciousness may evolve. That all may evolve and live from a vibrational state of collective soul mastery, where the greatest good of all is held as primary. This is why the choice to consciously live in soul mastery is so important.

When you are soul mastery in action, your actions are inspired from a higher source energy field. It is an energy field greater than you, and yet you are the vessel through which Its energy works. Its energy is always for the greater good of all, not just for self or for one segment of the population.

The wound is that place where the light enters.
Rumi

~

I choose to dance at the edge of mystery,
and unlearn all that I thought I knew.
I choose to purge, merge, integrate, and transcend
all perceived past and current realities
—physical, mental, emotional, and spiritual—
into higher dimensional awareness.
As I claim, embody, and anchor the Truth of who I AM.

~

Soul Mastery, is a conscious and progressive, energetic and vibrational absorption, assimilation, and integration process that ultimately creates a transformation of energy. It is not an intellectual, linear, or step-by-step methodology that requires you to grasp or memorize concepts.

In fact, do not try to grasp for understanding the through the intellect. Why? Because the intellects default, when it does not understand something, is to create resistance. And resistance, moves you out of vibrational alignment and holds you hostage to lower vibrational oriented mind-sets.

Thus, soul mastery invites you to remember that this is not a learning encounter, it is an unlearning process; unlearning of everything you thought you knew. It is an invitation to absorb, assimilate, integrate higher levels of vibration and frequency, to get comfortable with dancing at the edge of mystery, surrender, non-attachment, and non-resistance. As an unlearning process, soul mastery works on multiple levels of being, clearing energy through releasing more light into your cellular structures and body system.

In unlearning, you start to see that everything is energetically coming to you to show you who you are, and who you are not! You to begin to see everything that arises around you as an experience that is a part of an energetic and clearing process, and a soul-growth opportunity. Soul mastery, on a personal level, invites you to willingly and consciously dissolve and release the hold of your ego attachments, negative beliefs, negative perceptions, and hidden shadow aspects—which is huge—and may cause resistance to arise.

Therefore, if you sense that an idea, concept, or statement is—not resonating, or creating a sense of resistance—know that it is a signal to pay attention. Resistance, non-resonance, and not understanding are often subtle ways that the adverse ego/small self, uses the intellect to create confusion. The adverse ego, uses resistance, etc. as a show-stopper. It sets up a state of resistance, through confusion and not feeling good enough, in order to stop you from moving forward.

So, relax, take in the energies of the moment, lean into the energy and vibrations arising in the experience, and just observe everything, arising in and around you with a sense of "ahh, this is interesting!"

Know, that if you don't like what is appearing in your life, you can change it! No one else can do it for you! You get to choose the reality you want to experience and what you want to experience in the context of that reality. You get to choose whether your experiences support you in being a leading-edge consciousness or hold you hostage to the past.

You get to choose whether you create an environment where your desires can expand without triggering fear, resistance, a need for control and manipulation, or not.

Wholeness ... is not achieved by cutting of a portion of one's being,
But by integrating the contraries...
One does not become enlightened by imaging figures of light
but by making the darkness [the unconscious] conscious.
Carl Jung

~

It has been said that your true inner work begins when you are willing to begin to practice honest self-observation and confront the shadow aspects of yourself. Carl Jung, in his *Collected Works*, impresses the importance of making the darkness—the unconscious shadow aspects, the veils of illusions— conscious. And the journey to soul mastery cannot be effectively initiated until you are willing to say yes to engaging in honest and conscious transformation.

A simple spiritual practice such as, honest self-observation, and self-inquiry, can create a huge shift in the show-stopper type energies. I learned the importance of self-observation and self-inquiry from a life coach. During our coaching sessions, she would ask me to honestly observe the situation we were discussing, and to tell the truth regarding what I was seeing as my show-stopper. This spiritual practice made a huge difference in my journey of self-awareness, and soul evolution.

Honest self-observation, self-inquiry, and noticing are such simple tools, yet when practiced diligently, they initiate a transformation of dense shadow energy, etc. into a higher vibration. A vibrational energy that now works for you instead of against you.

The mind and body ... have power to transform energy
From one plane of consciousness [3ʳᵈ D., 4ᵗʰ D., 5ᵗʰ D....] to another.
This is the power and dominion implanted from the beginning...
And the climax is set forth in the resurrection and ascension
[into higher dimensions of awareness].[15]

~

Does it interest you that your experiences are the fodder for the spiritual-compost that creates your soul growth and evolution? As you are willing to practice self-observation and then assimilate the energetic richness of an experience; that experience becomes spiritual fodder that feeds your soul, and urges you forward on the evolutionary spiral of life.

The story that is about to be shared is one of those types of soul-feeding experiences. It was also, synchro-divinity, at its best, as it

---

[15] Charles Fillmore, *Revealing Word,* power; pg. 151.

resulted in an unplanned, unsolicited, spontaneous shadow-dissolving experience. It was an experience that spontaneously released eons and life-times of shadow-stuff from my subconscious mind and catapulted me forward on the evolutionary spiral. The following is what happens when the unified field of consciousness puts its possibilities into play.

It started with a moment of synchro-divinity, in the form of a gift. Someone, out-of-the-blue, walked up to me and gave me a book. The book was, *"Power vs. Force"* by David Hawkins, M.D. After reading it, I received an intuitive nudge, that I needed to go to Sedona, Arizona to attend a lecture and workshop by the author. I also knew that I had to make an appointment for a private session with Dr. Hawkins. Why, I had to do these things, I did not know, I just knew I had to do them.

The one-on-one session, for me, was an interesting composite of informality, self-observation, inquiry, simplicity, and clarity seasoned with a large helping of giggly. In fact, after our time together ended, I noticed that I felt quite "giddy." Not a term someone would necessarily use to describe me. I found Hawkins to be engaging, humorous, intelligent, eccentric, and a deeply spiritual being.

As I was leaving the compound, after the appointment, unexpectedly, a man came up behind me and tapped me on the shoulder. This interaction to my surprise, initiated the first of several synchro-divine events that would transpire over the next few days. The man shared that I might want to get to the retreat center early the next morning, as the seats upfront go quickly.

At the same instant, as the man was speaking, I heard my Inner Voice say to me, with great conviction; *"If there is only truth being expressed in that room and I AM That Truth, it does not matter where you sit—"I" will do what needs to be done—to you and through you— regardless of where you sit—Trust!"*

The clarity of this declaration resounded throughout my being and it shocked me. I felt like I had been stunned by an electric cattle prod, and I realized this was a sacred soul message meant only for me; the person next to me did not hear it. The next morning as I arrived at the retreat center, lo and behold, the man was correct. People were lined up very early to get a front row seat.

Remembering what my inner voice had placed upon my heart the day before, I consciously chose to not engage in the stampede for a front seat, instead I consciously made the choice to take a seat in the very last row, in the back of the auditorium.

As I listened to Dr. Hawkins share ideas on his "map of consciousness," one that particularly caught my attention was that of evolving from the level of conscious awareness of pride to courage. The levels of the map of consciousness are not a hierarchical model, but a transcend and include evolutionary model. The current level of conscious awareness is enfolded into the next evolutionary level rising; thus, the map is a descriptive listing of the rising and progressive evolutionary unfoldment of conscious awareness.

Something about the statement of the warrior moving from pride to courage drew me inward. I closed my eyes and was consciously practicing intentional-interior listening through the ears of my heart. I have found that the act of intentional-interior listening, as a spiritual practice, allows me to be fully present as it circumvents the words and goes to the "energetic-heart" of the words being spoken. It allows me to engage with the deeper energetic essence of the message rather than the messenger. In this state deep intentional-interior listening, I heard Dr. Hawkins voice share the following (the essence of my remembrance):

Marines enlist people on the level of pride, thus the logo, "Be a Marine." But in order to engage in battle or warfare, one needs more than pride—one has to move to the next level of awareness through— courage. It is pride that engages us to join in the cause, whatever that cause may be, but it is courage that allows us to do the things that need to be done once engaged in the cause. Thus, each of these evolutionary movements on the map of consciousness works towards honing the courage of interior warrior, who ultimately evolves into the archetypal spiritual warrior.

Why, do we need the spiritual warrior? The spiritual warrior is the warrior tempered by love, and thus, now knows how to be a force, without being forceful! And as enlightenment, is the ultimate goal of the spiritual path, it is imperative to perfect our spiritual warrior, for when we reach that final moment, when we are called to cross over the threshold of death, into the consciousness of enlightenment, it is our spiritual warrior and only our spiritual warrior—the representation of love—who takes us across.

One cannot reach enlightenment or nirvana, without facing the "death and annihilation of the ego." And it is the spiritual warrior's love and grace that allows us to face the "greatest enemy," which is the death of (adverse) ego, in whatever outer form it chooses to take, in order for our lesson to be complete. The spiritual warrior has the courage needed to walk through the veils of the "illusion" of death, betrayal, or annihilation, and come out on the other side.

The above words stirred something sacred in the interior recesses of my heart. When Dr. Hawkins made the comment on the spiritual warrior's role in enlightenment, in a nano-second—when Spirit is at work everything happens in a nano-second—I was embraced and enveloped by a light. It was a vibrating, luminous, brilliant, golden-white light.

Perhaps, it was the same light Saul experienced on the road to Damascus. For like Saul, hidden within the brilliance of the light was a revelatory experience. Before my interior eyes, I saw all the various ways I had engaged my untempered warrior in this lifetime. I saw all the people I had hurt, some intentionally and others unintentionally. And before I could blink, the entire history of my warrior-Samurai past lives, were displayed on the screen of my soul.

Undulating, waxing and waning, one chasing another, "feeling" pictures sped by. With each warrior, a feeling-picture clamored for my attention. The energetic sensation of the experience was that was I vacillating between feeling: a sense of surprise and shock at what I was perceiving; remorse and sadness for what I was capable of doing; mild indignation, for this cannot be me; quiet contentment for what was, and what now is.

As if all of this was not quite enough, suddenly, at the end of this multi-lifetime warrior picture show, I was catapulted to the plane above the corn field in Pennsylvania, on 9-11. Here, I stood face-to-face with the terrorist "warriors." Interiorly, I was asked by my Higher Self to view the picture without any judgments, to rise above the content of the image facing me—how horrible and how could they—and move into the context of the bigger spiritual arena, the evolutionary map of consciousness.

In the same instant, as I heard the request for non-judgement, my heart melted, I realized that these men—just as I had been shown that I had done over my own multiple life times—these men were playing out on the stage of their life-time. The were playing out a movement in the map of consciousness—playing out their own soul evolution.

It was overwhelming, there are no words to truly describe the context of the moment, as energetic waves of forgiveness and love washed over me. Suddenly, in a loud and booming voice, I heard the Inner Voice speak these words to me; *"Father, forgive them for they*

know not what they do." "Father forgive me, for I know not what I do."
I realized that I was being asked to engage in an act of forgiveness, greater than anything I had ever here-to-for imagined. I was being asked to move into a consciousness of unconditioned love.

I was being called to forgive, not from the sense of forgiving the acts, but from the sense of being willing to be a transparency that love as an unconditioned field of harmonic resonant energy could flow through. I was being asked to be unconditioned by any personal bias, cultural influence, societal influence, and more. I was being asked to be a transparency for love that can embrace all souls journeying through the morass of life–regardless of what they have done.

As I felt the energy of unconditioned love flowing through me, a soul wrenching, uncontrollable sob made its way up through my being and out of my mouth. The cry came forth like a spark that once ignited sets fire to a field of dry of brush, and once the fire gains control there is no stopping it, until it has completed its scorching deed.

I couldn't stop the sobbing; even if I would have wanted to I knew I couldn't, for the energy of the sob was releasing and dissolving shadow aspects—of shame, guilt, self-righteousness, need for control, pride, and so much more—that had held me hostage for life-times. At the same time, I knew the energy of the tears were a healing balm for my soul and for collective consciousness.

In a nano-second, I had the feeling that my warrior self was released, and my spiritual warrior had moved in, and I was being asked to rise above the pride of being embarrassed for sobbing, of being seen as emotionally out of control, of losing it; into allowing myself this moment of expressing unconditioned love for myself, and all others. I was being asked by the Universal Presence to dance at the edge of mystery, to participate fully in the exploration of the mystery at hand, and to be a transparency for what was desiring to make itself known to me.

As the intensity of my cry began to rise, an energy swept over my soul, and as a wind that sweeps in and clears all the little particles of dust and leaves that remains on the porch, in the fall, it swept away energetic remnants of past beliefs and perceptions. I realized that for each person that I was willing to "forgive," to embrace in the arms of unconditional love, in the context of the big picture of life and evolution, (regardless of whether I knew them or not) I too, was being forgiven for all the acts of perpetration that I had committed, in this lifetime and beyond. As I forgave, I was being forgiven.

To add another dimension to the experience, I noted that as I sat there crying, I was actually outside of myself observing the whole experience. It was a paradox, I was witnessing what was happening in real-time and yet, I was in the midst of a no-time, no-space experience.

As I started sobbing, the fire of Spirit spread, and others began to cry. Dr. Hawkins stopped his lecture, and said, *"Ahhh, the Holy Spirit has arrived, let us go into meditation."*

In the context of, I-never-saw-that-coming; often, when there is a movement of spirit, there will be a confirmation of some sort given. When I began to breathe again, a friend turned to me and said; *"Wow, I do not know what was going on with you, but as you began to cry, I turned to look at you. What I saw instead of you, was you as a Samurai, an ancient Oriental warrior!"* Yes! She used those exact words, a conformation from Spirit that what I had just experienced was real!

> *You have to be willing to give up control over how or*
> *when something is going to happen*
> *and allow a greater intelligence*
> *to begin creating synchronicities and*
> *serendipities equal to their creation."*
> Joe Dispenza
>
> ~

In choosing to be grateful for the experience in the moment, and in being willing to dance at the edge of mystery with the warrior archetype that arose, instead of running from it, I uncovered my spiritual warrior. The tempering by the spiritual warrior archetype, through the fire of love, shifted me into a greater understanding of the meaning of unconditioned love. It initiated me into setting an intention to be a conscious force for good in the world vs. a force for good for self.

By being willing to look, see, and tell the truth about my warrior archetype, I was able to see the underbelly, the shadow side of it, and transform it into something more useful in my life—I am a spiritual warrior—I am tempered by the fires of unconditioned love.

> *One of the problems connected with thinking about the*
> *fourth dimension is that some persons tend to think of the*
> *fourth dimension as the Absolute.*
> *It is not the Absolute, but only a dimension which transcends most of*
> *the current limitations of three-dimensional existence..."* [20]
>
> ~

Having reviewed what soul mastery is and how it reveals itself in your life; now it is time to ask the hard questions. What in your life—being, mind, and body—gets in the way of your conscious development of soul mastery? What is your show-stopper? What beliefs, shadows, perceptions, and veils of illusions continue to hold you hostage?

What are the—thoughts, ideas, reasons, feelings, emotions, spontaneous reactions, people, fears, pain, emotional residue, beliefs, conditioned perceptions, karmic influences, stories, veils of illusions, veils of forgetfulness, shadow stuff, dense energy, soul contract influences, personal narratives, and more—that come up and say no, I don't think so? You are not smart enough, good enough, rich enough... and so, you stop! You stop in your tracks and hold back out of fear!

Any and all of the above arise, typically, when you are up to something big, when you are desiring to reach for greater expansion of awareness, or have a desire to experience your potential, spiritual or otherwise. They remain in place, as personal show-stoppers, until you consciously make a choice to do something to shift and transform their energy.

Behold, I saw new heavens and a new earth,
and the old earth had passed away...
And there was no more death, sorrow, and pain.
Behold, I make all things new.
Rev 21: 1,4,5
~

Know that it is your adverse ego that keeps you stuck in the muck and mire of life when you could be soaring in the heavens of a new reality. It is the adverse ego that wants what it wants, that feels righteous indignation, and that feels justified in fighting for things or using greed and manipulation to get whatever is deemed justified.

The transforming and clearing process of soul evolution invites a letting go of all of fears "energetic relatives;" anger, anxiety, control, lust, labeling, greed, addiction, judgment, righteousness, manipulation, a need to "feel" safe, unconscious narcissistic tendencies, a negative sense of identity, environmental fears, a need to be right, a need for power, playing small, to be haughty, to be seen as powerful, acting better than, resistance, etc.

It is these types of third-dimension oriented, dense-fear-survival-adverse ego identities that you have said yes, to transforming. And they can be transformed as they are brought into the light of honest, conscious awareness.

Enlightenment is the crumbling away of untruth.
It's seeing through the façade of pretense.
It's the complete eradication of
Everything we imagine to be true.[16]

~

One of the shadow-emotions that I danced with for the first three decades of my life, both unconsciously and unhealthfully, was fear. Fear that I was not enough and that there was not enough. Not enough money, not enough time, and not enough love for me. This survival emotion revealed itself in me through subtle behaviors—or perhaps not so subtle to those could see through them.

Behaviors such as aloofness, coolness, wanting to be one of the smartest people in a room, insecurity compensated by overachievement, and more. Did I see these? No! I thought it was everybody else who was strange, who didn't understand me, who didn't like me, appreciate me, and those fear-oriented thoughts and feelings provided me with reinforcement and proof of my unconscious sense of unworthiness.

It took until I was in my thirties to discover, I was enough! I had been living in emotional resistance to the truth of my being, and that resistance created a <u>sense</u> of disconnection, in me. I was not actually disconnected from Truth or Spirit. You can never be disconnected from Spirit; you can only have the sense of disconnection by allowing something else *to seem bigger and truer* than the truth of the I AM, the true Self.

After I had been in New Thought for just a few years, a long-time Unity minister said to me, "Don't think that a pretty face or being smart is going to do you any good on the spiritual journey. Let them all go, be honest, and discover who you really are!" I had no idea what she was talking about, but her words haunted me.

Her words, and the truth energy held in her words, triggered an unconscious desire to know what she was talking about. That desire, unconsciously, called forth my first unraveling experience, and bout of true self-honesty. It was an invitation to see through the shadow-energy, the veils of illusions, and veils of forgetfulness that were residing in me.

During the first few days of the unraveling experience, the first of many to come, I made her what was wrong with me. I thought over and over, "How dare she say that to me? She doesn't even know me!" The unraveling of energy wasn't pretty, but I see now, it was required. It

---

[16] Adyashanti, *Resurrecting Jesus: Embodying the Spirit of a Revolutionary Mystic*, Amazon

was time that my inner five-year-old, with its limited beliefs and perceptions, stopped driving my bus, and directing my life.

As it was with me, in this story, know that your human-ego-survival-oriented reality is created from old earth, third-dimensional realities that are illusionary. Illusionary in that they are created from the perceptions and emotional residue of past experiences and live only in your mind.

Only you can go back into story and retell and relive it over and over and embed your emotional residue deeper into the cellular memory of your body and mind. And, in the same vein of thought, only you can say yes, to honest self-observation and conscious transformation.

Vulnerability is the birthplace of love, belonging, joy...courage, empathy, and creativity...accountability, and authenticity. If we want greater clarity...and meaningful spiritual lives, vulnerability is the path. [17]

~

Emotions! What are they good for? From a third- dimensional perspective, emotions are a trigger that moves you into being reactionary and/or emotionally out of control. From a soul mastery, evolutionary perspective, emotions are teachers and evolvers of consciousness. You are designed to evolve, and emotions are designed to be informers and instructors for that evolution.

Emotions can be healthy as you learn to express them constructively and overcome their hold as meaning-making, reaction-sparking mechanisms of response. Emotions can be unhealthy if they are suppressed, repressed, or expressed in a critical or accusatory manner, and played over and over in your mind.

Over time, emotions have been misused, mis-created, and oriented toward survival reactions. Additionally, they have become perception-creators, meaning-makers, and creators of instantaneous reactions from the meanings you have made from them.

Emotions are energy that hold a frequency and each type of emotion—fear, anger, sadness, jealousy, greed, joy, appreciation, trust—vibrates at a certain frequency level. Joy, appreciation, trust, etc., vibrate at a higher level of frequency than greed, anxiety, and jealousy.

---

[17]Brene'Brown, <u>Daring Greatly: How the Courage to Be Vulnerable Transforms the Way We Live, Love, Parent, and Lead</u>

Emotions were created to support you in seeing and feeling what is happening in the moment.

However, if emotional responses are not utilized for the purpose intended, for instance when they are repressed. Or, if they are not dealt with in the moment in a constructive manner, they have the capacity to become embedded in our cellular memories—especially if this is a practiced response—and this creates, overtime, shadow energy, dense energy, veils of illusion, and more.

For example, you are a seven-year-old, innocently doing something, like drying dishes. Your carefree little being was happy doing what it was doing when out of the blue, your mom or dad came in and started yelling at you for not doing it right. Perhaps they even told you that you were stupid for doing it that way.

As a child, you knew you wanted to respond or even cry, but you were afraid. So, you repressed or stifled your emotion. Once doesn't necessarily hurt you, but if this becomes a pattern of getting yelled at and then stifling your emotions, that emotion-stifling pattern gets embodied and creates dense-energy, known as the pain-body, shadow-energy, veils of illusion, etc. And unbeknownst to you, these energies, then, unconsciously, run your life.

Then, later on in life, someone at work mentions that perhaps your report might have been better if you added a resource section, and interiorly you go crazy or perhaps, you even start yelling at them. You have no idea where the reaction came from, but there it is! This is not about right or wrong or judgment; it is about what has been created by you. You are not alone in this.

Once an emotion is expressed constructively, you return quickly to harmonic resonance within your body. When emotions are felt, acknowledged, and expressed in the moment, it provides you with the opportunity to rebalance and recalibrate the body system. When emotions are not expressed, they get repressed into the body system and they create resistance, which shows up in all types of ways in your life.

"Enlightenment is a destructive process." [18]

~

Crisis precedes transformation,
but not all crisis leads to transformation. It is a choice!
You are the architects of your next step of evolution!

---

[18] Adyashanti, *Resurrecting Jesus: Embodying the Spirit of a Revolutionary Mystic*, Amazon

The evolution of consciousness is the driving force behind all physical evolution. Not one thing in your life occurs without purpose. [19]

~

An important brain research discovery, that substantiates the role of the subconscious mind, energy, and emotions in soul growth; and being discussed by today's leading-edge teachers and researchers (Joe Dispenza, Eckart Tolle, Mark Waldman, Bruce Lipton, and more), has to do with the science and mechanics behind how and why people hold emotional residue in the body.

Researchers and scientists have labeled this subconscious, unconscious, emotional, residue holding-pattern; the pain-body, drama-body, and emotional-body. Regardless of the name or label you give it, from a scientific perspective, it is a stimulation and reaction effect that arises in the brain, from an unconscious neuro-chemical response mechanism.

The stimulation and reaction effect elicit a feeling of an energy that feels familiar and almost comforting, and its continuation over time becomes emotionally addictive. Your reaction becomes a known feeling that is fed by an ever-growing addiction to the neurochemical release that occurs when you get upset about a memory or something that has happened in the past. Have you ever met someone and instantly disliked them but did not know why? Then later a memory surfaces and you recognize that the person you just met looks like your Uncle Joe, who is nasty.

Over time, memories become unconscious and get embedded in your cellular body system as cellular memories. When you are functioning unconsciously from memory that is steeped in lower survival emotions (and embodied cellularly), it can be recalled in a millisecond. This unconscious emotional hostage-taking is instantaneous, and it invites you to go deeper into your storytelling, which only serves to anchor your story deeper into your physical being and then uses your energy to stay in limitation. When you decide to make a change, memory cells stored in the body created from embodied emotional residue can't accept the idea of change (think automatic survival mechanisms). The change is often too much of a stressor and your own memory cells work to sabotage you.

They stop you from moving forward, and this happens over and over. The sabotage is subtle and often unrecognized, but it stops you,

---

[19] Robert Brumet, *Finding Your Self in Transition,* Unity Publishing.

nevertheless. It can be a thought or feeling—*just give up, this isn't going to happen, you are not going to do this or get this done*—and you stop. Now the body, emotional pain body, is in control and not the brain.

However, all is not hopeless. Consistently thinking and feeling in new ways (such as engaging empowering questions, self-observation, meditation, etc.) invites the creation of a new norm. By sensing the future being felt in the now, over time this future feeling in the now causes an effect in now.

From a quantum perspective, you are consciously *causing effect* rather than trying to predict and control effect. You cause effect through consistent repetition of feeling a thing as if it is happening now, as if it is already complete, and this changes your body signals. Then these new signals create a change in your neural patterning and in your neurochemical responses, and that shifts you into new levels of energy, vibration, and frequency.

> *Why would I believe fear, when it is such a liar?*
> *Fear, what is it good for? Absolutely, nothing!*
> ~

Resistance is created when emotions and feelings, the end-products of your life experiences, are repressed instead of expressed. Resistance interferes (inner-fears) with resolve and resilience, and it takes resolve and resilience to bounce back from perceived adversity, and to overcome the effect of resistance and negative emotions; and to halt the development of the emotional pain-body.

When you choose to change your mind regarding your experiences, your reality changes. With resilience, and a soul mastery mindset, you can develop a spiritual mind discipline, which supports and allows you to honestly observe resistance within and around you.

Regardless of where the resistance is arising from, you have the capacity to relax, as you realize that resistance, in any form – fear, anger, upset, jealousy, wanting to be seen as right, etc. — has only the power you assign to it. In this recognition, you can choose to shift your awareness and invite your higher soul self to call forth soul-utions.

As you claim your I AM dominion, authority, and sovereignty as who you are, the universe will reflect back to you, in all areas of your life, in alignment with that claim. Claiming the dominion and power of the I AM, without reservation, resets your vibrational set-point. Resetting your energy, vibration, and frequency shifts you into a higher dimensional level of awareness of who you are.

In this higher state of awareness, worldly things are no longer valued over soul purpose, for it holds new qualities that you begin to naturally live from. Qualities such as appreciation, respect of self, honor, trust, value, compassion, peace, collaboration, cooperation, love intentionality, clarity, etc.

To change something, you must offer a vibration that is different from that which is occurring in the moment. So, observe when thoughts or feelings of dissatisfaction or negativity begin to arise and want to be spoken into and make a conscious choice to halt any continued progress of the negative-oriented thoughts or feelings.

Choose to engage the spiritual mind discipline required to disconnect from the negative energy before it gathers momentum. In the beginning, this will take concerted and directed effort. However, after a while you will find that you are gaining mastery over your reactions, thoughts, emotions, and feelings.

You are a facilitator of energy, vibration, and frequency and everyone—consciously or unconsciously—is facilitating energy in each moment, and in every experience and interaction. For everything is energy; your thoughts, feelings, perceptions, and even your body, it just looks solid. The more elevated your vibration is, the more capacity you have to be a conscious alchemical transformer of energy in the moment, and to shift any negative energy that is arising.

You can choose courage, or you can choose comfort,
but you can't have both. Not at the same time." [20]

~

"You either walk inside your story and own it
or you stand outside your story
and hustle for your worthiness." [21]

~

Life is an ever-progressive, ever-expanding, ever-evolving, upward spiral of conscious evolution! If this is true, and it is, then what if the possibility exists that every person and situation you have encountered, has said yes, on a soul level, to be a part of your soul narrative? What if you came here to earth with a soul contract in place, that included all of the major players in your soul narrative? What if, each player agreed to participate in some way, in supporting you in your

---

[20] Brene' Brown, *Braving the Wilderness:* The Quest for True Belonging and the Courage to Stand Alone
[21] Brene' Brown, *The Power of Vulnerability*, TED Talk

soul growth and evolution? Remembering, soul growth does not always look pretty!

If you truly knew that every person you ever interacted with in this third-dimensional experience has said yes to be a part of your soul dynamic and adding information to who you are—how would it change your story? How could it shift it into a strength-oriented narrative?

There is a difference in the way narrative and story are being used here. Most of the time when someone says, "let me tell you my story," their story is associated with a negative scenario that has defined them and held them hostage to victim-type energy. They are not over it, not healed. A narrative is a story in which the wounds and the shadows have been healed, and can now be shared without any energetic attachment, except as a lesson learned.

Your mom, your dad, your friends, your experiences, all of them—what if they were a part of a soul contract that was meant to support your soul transformation—how would that change how you view those relationships? And, what if, that soul contract was designed to influence and inform the type of experiences you would experience in life? Experiences designed in order that you might evolve and expand your soul awareness—or not—in accordance with the choices you made?

How would knowing this shift your perspective on the experiences of your life and the story you tell about your experiences? How might it shift how you see and interact with people, or with your family? You create either a story or a narrative. It is your choice which one you choose, and shape through your beliefs and perceptions.

Your soul contract isn't meant to hold you back, throw you into a tizzy, or hold you hostage; it is meant to ensure your success. Yes, success! Yet, it is up to how you choose to perceive the elements of your life, and then how you choose to share them? Story or narrative? Many people, twenty or thirty years after leaving home, are still singing the same song; "I am not who I could be because—my mom was this, my dad was that, you just don't know, I will never get over it." That's a story!

Consider this, what if Judas and Jesus had a soul contract? What if, what looked like a great betrayal for a few silver coins was really a part of a soul contract agreement? Aren't we all betrayed by someone we trusted, at some time? What if, betrayal is part of the soul contract? Could the crucifixion have taken place without Judas' agreement to betray Jesus? In the Gospel narrative of Gethsemane, does it say anywhere, "Then Jesus whined and complained about how unfair Judas treated him; how he betrayed him?" How would it change you if you knew that you had a soul contract with all the main characters in your life?

What if, your contract has beginnings and endings written into it, also? What if, those ending experiences like divorce, getting fired from of a job, or the death of a loved one was the ending of a piece of your soul contract—because you were energetically complete—ready for a new experience or phase of life—even if the experience didn't appear to feel like you were ready? How would these ideas, if you believed them, reshape the perception of your story and all of your myriad of experiences? What new meanings could be created from the context of your experiences?

Please don't consider this predestination—it is not. You, through your conscious mind, always have freewill. You get to make a choice, to decide, what you will glean from the experiences of your life. Here is a personal example.

My mother was a paranoid schizophrenic and from the age of ten, about every six months, my mom left home to go to psychiatric institutions for months at time. Which meant that, as the eldest child, I was in charge of the family: the cooking, the cleaning, household chores, taking care siblings, etc. After a short period of time, even as a child, I grew bitter. I began to to believe that my mom came to earth to make my life miserable, and I hated her for what she did to me. I chose to live with that perception for twenty years, and it affected my life. I was miserable. And my life reflected that misery, over and over. And it was all because of her—it was her fault—and I let everyone know it.

When I found New Thought, and the teachings on forgiveness, my awareness began to shift. I realized that I was the one carrying the ugly energy, not her. She loved me, she told me that, over and over. It was me who went out of her way to ignore my mom, and be ugly to her. This went on for years, until I was at a Thanksgiving retreat, at Unity Village. During a meditation experience, my heart burst open, and love and light flooded my being. And my mom, came into the midst of the light, as a beautiful young child, and I saw who she really was.

In that moment, I stopped hating her and feeling like she ruined my life. I realized that she was dealing with what life had given her, and that she was the one who really had a hard life, not me. I went to her and asked if we could start over? She said, *yes, of course my darling*! And never was another word spoken about all things that happened between us. My heart was open, I was forgiven! Not by her, but by me!

Years later, after the luminous event, I was reflecting on that meditation and forgiveness experience, and my understanding of what occurred, shifted and rose to new level of awareness. It spurred to ask new questions: *What if* my mom and I had a soul contract? *What if* she agreed to come to earth and take on the appearance of a paranoid schizophrenic so I could have the life experiences that would shape me into the person I am today? This thought shocked me!

As I began to perceive that new awareness, as a possibility in my field of reality, the scope and landscape of my current reality around forgiveness shifted. I saw my mother with an even more loving heart and softer eyes. I saw our relationship differently. I saw her as someone who said yes to a physical and mental condition in her life, in order to play a part in my life, through a soul agreement. She did it to support me in my soul growth and ultimately into the journey to soul mastery. Wow, what I had previously seen as an act of betrayal that ruined my life—shifted and transformed—and I saw the true selflessness in what she accepted for her life experience!

Our soul contract shifted the moment I saw her in that new way. She became a true friend, no longer an enemy who betrayed me. And more than anything, I realized this wasn't about her; it was about me and the story I had made up to serve a purpose in my life. When the story alchemically transformed, so did I. I was truly free!

I realized that soul contracts and their elements serve a purpose. When you are complete with various elements of a soul agreement, the person and/or situation is released, and new dynamics arise to support you in ascension of awareness, of ascending into the next phase of your consciousness adventure.

The cells of the body are centers of force
in a field of universal energy ...
That which appears solid is in reality the scene of constant activity.
The eye is not keyed to the pulsations of
Universal Energy and is deceived...
All energy and life are governed by laws of spiritual harmony...
If the mind that receives sound vibrations
is in spiritual consciousness, the body responds and
breaks up crystallized thoughts. [22]

~

---

[22] Charles Fillmore, *Jesus Christ Heals*, pg. 172

What does it mean "to break up crystallized thoughts"? It is the breaking down and releasing of old structures of perceptual programming, held within your mind's neural, and your body's cellular, "memories". These are the memories that have been, over time, and with continual reinforcement, embedded as conditioned emotional patterns and responses.

Scientific research reveals that the strands of your DNA embedded within your cells, and cellular structure, are encoded with a massive amount of data and information. It is said that one gram of DNA can potentially hold up to 455 exabytes of data. Yet only a small percentage (less than 5-10%) of that inherent information has been decoded by the scientific community.

Would it excite you if there was a possibility that the remaining information held within those strands of DNA pertained to higher levels of awareness and truth? Perhaps humankind is finally ready to have the full knowledge of the patterns concealed in the makeup of our DNA revealed and brought into the light? There is much that the scientific community has discovered, and there is much more to understand!

Soul mastery, and spiritual growth, is predicated upon the idea that to evolve you must take on new levels of light, information, and knowledge, such as that found in higher awareness and frequencies. However, that light is not something you take on, go after, or get. That light, when all is said and done, is something that you release from within the hidden recesses of your DNA and cellular structure. It is already there within you, it just requires a cellular shifting, rebooting, or change to release it. This change or rebooting is often referred to as a DNA restructuring.

Quantum science also refers to this idea of the impact of the release of cellular light and DNA restructuring on the body system through its explanation of epigenetics. Simplistically stated, epigenetics is referred to as the impact or amount of light in a cell.

Research studies have shown that the higher you raise the amplification or level of light in the cells through meditation, joy, trust, appreciation, etc., the healthier and more whole you are and stay. The dimmer the light is, created by stress, resistance, negative thoughts, and out-of-control emotions, the greater is the chance for illness to appear. In each scenario, quantum and metaphysical, the amount of light in the cells creates changes in the cellular structures and initiates a rewriting and restructuring of your cellular codes. Metaphysics and metaphysicians have said for ages that every cell of your being has substance, life, and light/ intelligence. As you release old patterns from your subconscious—cellular memory mind and body system—you allow a greater amount of the light of higher source awareness (already contained and embedded within the substance of your cellular system) to shine through your being.

Embedded, crystalized, and conditioned memories often elicit negative emotional responses, often in a nanosecond, especially those that trigger a feeling of familiarity. The ego energetically feeds on the emotional residue, and, thus, wants to keep them alive. When you consciously make a choice to release crystallized energies, or if a spontaneous releasing occurs, it opens the way for more light and energy to be present and others notice!

Kundalini has been defined as, 'She Who is Hidden.'
Kundalini, a divine feminine energy, lives at the base of your
spine and lies dormant until spontaneously awakened.
When Kundalini energy awakens,
an alchemical transformation occurs within your being.

~

In the scientific world, as mentioned, the releasing of cellular light and energy is referred to as epigenetics, or DNA "restructuring". In the spiritual world, the catalyst for the cleansing of these crystallized memories, is known as Kundalini awakening, or Kundalini rising.

Kundalini is said to be a serpent-like energy that lies dormant within energetic body centers or chakras. Once this energy is awakened, it rises and begins to move throughout the entire energetic body system. Its movement initiates a systemic ego-cleansing and clearing, old-program-unwinding, and perception-distortion-unraveling that creates the conditions for breakthroughs into the next levels of spiritual growth. However, this is not an overnight process.

Through this Kundalini work, the light that lives within your DNA and cells is released and DNA "restructuring" is initiated. The DNA restructuring creates a greater capacity for the holding of higher levels of light frequency, within the body system.

The restructure of the cellular structure system changes everything and, ultimately, creates a capacity for new values to be held in consciousness. For example, what has previously been held has a value for a need to be safe and feel protected will be shifted into a new value that hold a consciousness of unity and collaboration. An alchemical transformation occurs, and you move out of the ego-distortion emotional need for safety and protection—which is a survival/root charka orientation into a new value system that holds higher levels of awareness toward unity consciousness and the integration of light.

Your body, then responds to these higher vibrations by restructuring, dissolving, and releasing old encoded and embedded cellular messages, memories, and reactions, behavioral and emotional, that are automatically triggered by certain types of situations and people. Thus, you are participating in DNA restructuring all the time, as you are willing to hold a higher awareness through meditation and as you are willing to change.

It is also important to mention that the greater amount of resistance you hold within your body system, the more you fight the emergence of the light. Holding onto resistance, especially during a detox, cleansing, or change experience, results in increasing the amount of pain and discord you will experience in the midst of an experience.

Also, as light is activating cellular DNA restructuring, the change it creates has the capacity to create a new relationship with your body. Your body will start to want do things differently. You may choose to release certain foods, alcohol, sweets, or take on new eating habits. There are those that say that if you were spiritual, you would not eat that or drink this or that. The reality is that each person is guided by their own innate higher power, and thus knows what their body needs for each phase of soul evolution. So go with the flow!

I have cast fire upon the world,
and look, I'm guarding it until it blazes. [23]

~

Karma dissolves, as you evolve. Karma, as a third-dimensional reality concept is often referred to as, a dense energy that accumulates as a result of negative-oriented actions taken. Actions that often arise from conscious, unconscious and/or unskilled negative behavioral and emotional programming. Actions that are not in alignment with your Inner Source.

Karmic density is the residue or negative effect of those negative-oriented actions taken. Karmic density holds a magnetic-like quality that has the capacity to attract back to you the negative energy that you gave out. It is like a boomerang. You get back energetically, in the form of life experiences, what you have given out through an accumulative effect of your past actions and behaviors. Perhaps soul contracts are one

---

[23] Gospel of Thomas, *The Nag Hammadi Library, Logion 10.*

of the ways that you have agreed to clear your karmic density. Note, you can accumulate good Karma.

As you evolve into unity consciousness and higher fourth- and fifth-dimension awareness, you begin to unify all aspects of your being – spiritual, mental, emotional, and physical. As you do this, you release the dense effects of karma. Unity consciousness as a state of being is a unified state of being that has released the density that was created through duality thinking.

When you are in unity consciousness, you respect yourself and others enough to say, with clarity, *yes* or *no more,* to experiences and action that do not align with your new level of vibration, and to not to feel guilty about saying it. In unity consciousness there are no more unconscious acts or actions, no pretending, no fear, and no anger. You speak your words authentically, with courage, with clarity and intentionality, and with a knowing that you are holding the highest and best for all involved. The more you are willing to be in integrity with what you believe, be authentic, and have respect and honor for yourself (especially regarding how you want to be treated), others energetically recognize this shift in you and their dysfunctional behavior toward you stops.

The more you stand in your power and authenticity, the faster you collapse the karmic effects being held in your energetic field. The more you step into your authentic self, the clearer you are, the more you rise into union interiorly with all aspects of who you are: the faster the people that were entangled in your web of karmic energy will fall away from your life or shift their attitudes and behaviors toward you. Be ready!

A&A: *Soul Mastery - SOUL-U-tions*

*Anyone who develops spiritual power may enter
the book of Life within Cosmic Mind and read out of its pages.* [24]

As you live as an expression of soul mastery awareness, you live from a realization that knows–that it knows–I am the -U- in SOUL-U-tions and the following "living" truths are aligned with my current reality:

1. A multidimensional "living" reality, underpinned by the understanding of how energy and vibration works is—my new norm.
2. I live in a vibrational universe, and all my interactions are energy in motion, vibrational in nature—regardless of appearance! Energy doesn't lie, and I can't hide! No matter how well I think I can!
3. I am a spiritual vibrational being in a human experience, and as a vibrational being, I am a vibrational transmitter, a broadcast signal, for the attraction of my experiences.
4. As a vibrational transmitter, all the broadcast signals I emit are sent forth in vibrational alignment with the momentum of the energy that created them—in alignment with who I am! I am aware of the level of dimension and level of energy I vibrate!
5. No dimensional level of awareness—3rd, 4th, 5th, higher—is better than another, just different. I have the capacity to consciously choose the dimension I want to engage & experience from.
6. I am here to be an example, to be a way-shower into higher dimensional awareness. However, to be this example, I must first overcome my own self-imposed limitations and perceptions.
7. Everything that occurs in my life is a transmission of energy and vibration, and an alignment and activation experience.
8. All experiences *only* come to teach me about what I want—and don't want to experience in my life. They come to reveal who I am in each moment. Therefore, I do not get attached to what is happening, what I believe, or give meaning. It will all shift!
9. Transformation—of my past, my beliefs, my perceptions—is a transcend and include experience — not an eliminate and destroy.
10. My higher self/superconscious is not awakening, it is awake. It is my human self, with its accumulation of dense energy, that is re-awakening and remembering, the truth my soul holds as to—who I AM—and as a master facilitator of energy!
11. I recognize and consciously claim the dominion of the I AM/Spirit/Universal Intelligence as my soul awareness. I align continually with this claim. This is not an intellectual pursuit; it is an absorption and assimilation process. I know this from a heart/feeling level

---

[24] Charles Fillmore, Talk, July 5, 1929

and increase the velocity of my desire to know, claim, and being harmonic resonance with my I AM Self!

12. I pay attention, observe, and notice the things happening in my life— for they are mirrors that reveal where I am focusing my attention— and they reflect what areas of my life and being are calling for shifts in my thoughts, feelings, and emotions.

13. I work on myself-first. I utilize the tools of honest self-observation, meditation, never-the-less I Am willing, conscious engagement of spiritual mind discipline, forgiveness, gratitude, transparency, authenticity, vulnerability, energy-vibration-frequency, and...

14. I am willing to consciously work on healing the emotional body (my shadow aspects)—it is an imperative. The emotional body holds the remnants of the pain of my perceived story—the neglect, abuse, co-dependency, and loss—I perceived I experienced. Neither fear, nor pain cannot be taken into higher dimensions—they don't resonate!

15. My shadow aspects live in recurring patterns! I am alert to and observe with honesty the recurring patterns in my life and I am willing to "mine" them for the hidden messages they contain.

16. Practice, practice, practice: through practice I learn how to sense and interpret the subtle flow of information that inner guidance-inner source is sharing in every moment.

17. In every moment I make the choice to be intentional, focused, conscious, present, and aligned.

18. I build my capacity to observe and notice the energy present in each moment and in each interaction. I develop this as a competency.

19. I consciously ask empowering questions: *What would a person who is _______ (abundant, successful, etc.) be feeling right now?* This consciously brings my future desires into the now—and as I experience and feel them—as if already happening—now—I am Causing Effect—I am pulling them from the future into now, as already complete. This is different from the Law of Cause and Effect. I consciously release attachments and resistance by asking: *What meaning am I giving this? Is what I am about to say or do going to create separation, or invite wholeness?* The answer will let me know where my ego is still attached and holding on. I consciously choose to disengage from third-dimensional patterning and pettiness.

20. I am willing to consciously work on clearing and then integrating my mental-physical-emotional-spiritual bodies.
 In the fifth dimension, they are all one: they are in unity, in sacred union!

21. I consciously learn how to demonstrate a spiritual principle in mind (and I have the capacity to repeat it on command).

22. I know that the sense of too many obstacles in my way, feeling triggered, feeling stuck, or feeling overwhelmed is an energetic projection I have created from old programs, beliefs, and perceptions.

These energetic projections form an appearance in my reality that I have named— obstacle, triggered, stuck, overwhelmed, etc.—they feel real—yet, they are actually an illusion—for they can be changed. They are, in actuality, indicators from my emotional system signaling and informing me that there is new information I am ready to receive, or a new way of being that I am ready to grow into. It is my choice as to whether I choose to invite this new information or growth in, or not.

23. I get to choose, in every moment, what type of thoughts and emotions to hold in mind and body.

24. Life is an ever-progressive, ever-evolving, upward spiral of conscious soul evolution. I treat it as such!

25. If I require support, I will find a spiritual mentor or coach!

26. I am willing to heal my religious past; it is an action that will pay great energetic dividends.

27. The spiritual path has initiations, benchmarks, signposts, and more. As I discover what they are, it makes the path "easier" to traverse.

28. Spiritual principles (abundance, harmony, joy, peace, conscious neutrality, etc.), are mine to use, as I consciously align with them. Once demonstrated in mind, they are repeatable and sustainable, forever!

29. I choose to practice micro-meditations—presence pauses—over and over, throughout the day! Several times a day, I pause, and take a breath. I consciously, breathe into my heart space, bringing my attention in my heart space and then, BE still. I consciously create moments of energetic spaciousness that allow a connection with the true Self to spontaneously occur. This supports, in the moment, the capacity to engage conscious neutrality, and spontaneous right action!

30. I develop conscious neutrality, for it is a natural fourth-dimensional state of being that, by its very nature, invites a more highly developed and natural response to situations, especially, perceived discomforts.

31. As I _be_ the state of being of—conscious neutrality—I am a catalyst for the possibility of alchemical transformation to occur in every moment, and in every situation! Yes, I have that much power!

32. I expect and honor synchro-divinity experiences and their spiritual impact

33. I consciously, and frequently, pause to invite the presence of my higher power into my awareness! I know It is always there to serve as a guide, mentor, way-shower, and more. However, I must invite It in as an active participant in my life, as It will not interfere with my soul choices!

3

Everything is Energy, Vibration, and Frequency

*If you want to know how the Universe works
think in terms of energy, vibration, and frequency.*
Nicola Tesla

~

Everything is energy, frequency, and vibration! And how easily you see into the energy fields swirling around you, depends upon how heavy your veils of forgetfulness are! This next piece is about seeing through the veils of forgetfulness; the veils of illusion.

I was about ten years old, when my friend and I decided to go see her grandma; which required a long walk through the woods. I remember that even at that young age, I loved the quiet of the woods, the sense of being a part of nature. In this case, I also recall the sense of safety I felt, amid the woods, because I was with my friend. A short distance from her grandma's house, we saw her grandma's best friend and neighbor, outside in her garden. We were very surprised to see her, because she had been very ill the past few months and was bed-ridden.

She was looking healthy, standing erect, with a babushka on her head, and a hoe in her hand. She was standing amidst her beloved garden. We saw her, she saw us, and she waved. We waved back, and then she turned, and walked into the woods. It appeared that as she walked into the woods, she became invisible.

My friend and I looked at each other, shrugged, and smiled; for even, at that young age, we thought it strange that she walked into the woods. Having arrived at my friend's grandma's, we mentioned to her that we saw her neighbor, that she waved at us, and then walked into the woods. Her grandma looked at us strangely and said, "Girls, stop talking like that, you couldn't have seen her. She died this morning!"

Am I certain we both saw her? Yes! Can I explain what happened? Perhaps! I believe my friend and I, as young children were open, pure energetic conduits, the veils that often hold the energy of higher dimensional awareness at bay, were still thin. The dense energy of third dimensional experience had not sufficiently clogged our systems, or created impenetrable veils of forgetfulness, as of yet. Thus, we could see her energetic light body, which is invisible to most adults, for the veils are too thick to see through!

"Everything is energy and that's all there is to it."

~

Soul mastery, is not only grounded in spiritual principle, but scientific principle, also. Consistent and habitual engagement of spiritual and scientific principles and higher-level choices, overtime, begins to rewire the neural pathways of your brain. Once wired into your neural pathways, mind, and consciousness these principles are demonstrable, sustainable, repeatable, accessible, and permanent.

Here is a relatively simple overview of the latest in brain science and research that has a direct relationship to the soul mastery experience. Understanding the impact these scientific premises have on the demonstration of spiritual principle, will support you in your process of soul evolution, and your subsequent movement into mastery.

What syncs in the brain, links in the brain.

~

It is said that if you grasp the meaning behind the following statement, *"Neurons that fire together, wire together,"* you will have an understanding that can support you in the creation of clear and conscious choices and dynamic decisions.

Why, is this of particular importance to spiritual growth and evolution? Because, your choices and decisions affect the creation and development of the landscape of your current reality and reveal how "alive" spiritual principles are in your life—or not. *"Neurons that fire together, wire together,"* is often said, or referred to, in a variety of ways, some of which include: *"What wires together, fires together." "Synapses that fire together, wire together." "Thoughts that fire together, wire together." "What syncs in the brain, links in the brain."*

Regardless of which statement you choose to use, they all hold a foundational underpinning of both, scientific and spiritual principles. The spiritual principles are, *"Your words have power to create." "Thoughts held in mind, produce in the outer after their kind."* The scientifically proven principle is, *neurons that fire together, wire together.* What this means means from a spiritual perspective is—when you consciously decide to make a choice to change and you choose to engage a new a thought, feeling, belief, behavior, etc. (neurons firing together)—with continuous repetition—it has the capacity to rewire the neural pathways of your brain (wire together)—which will create a transformation of being.

What this means for you is that through your choices, you create the neural patterns, the paths of neural response, which create the landscape of your reality, which determines what shows up in your life, how you live your life, and the type of experiences that you will draw to you in life. Your choice of focus establishes the thoughts, feelings, beliefs, perceptions, or principles that become a part of the landscape of your awareness. You, and only you, have the capacity to create change and transformation in your life—or not—depending upon the choices you make in regard to the level of (3rd—4th—5th dimension) thought, belief, perception, or principle you choose to wire from.

The following, pictorial example of a neural pathway and its description, is meant to reveal on a simplistic and rudimentary scale, what the process of rewiring your neural pathways entails.

Example of a Neural Pathway

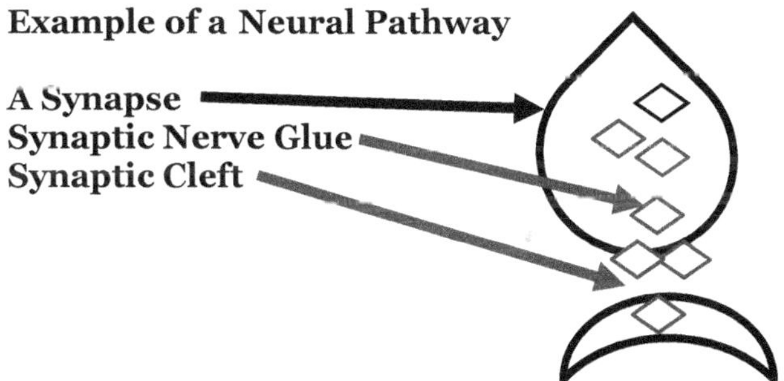

The brain is made up of collections of neural pathways that contain, among many things, synapses. Between synapses is empty space. That empty space is known as the synaptic cleft. When a thought arises, a synapse sends a chemical signal across the synaptic cleft to another synapse.

In the creation of a new neural pathway or for recall, this happens multiple times over many synapses. When the chemical signal goes across the synaptic cleft, it builds a bridge over which an electric signal can cross. That electric signal carries a charge, which is full of information that is related to and relevant to whatever it is you are thinking about or holding as thought.

Have you ever had a pain in your elbow? How did you know you had pain in your elbow? Nerves that go from the elbow to the brain carry electric signals-sparks-charges created from the elbow sensation point to the brain. It is in the brain then, that the signal is actually felt.

Each time an electrical signal is triggered, through repetition, the synapses grow closer together to shorten the distance the electrical charge must cross. Over time, with continual repetition, new neural pathways are created. This is what happens regardless of whether it is a pain signal being triggered or, the creation of a thought, feeling, or belief.

As the brain is rewiring its circuitry—*what wires together, fires together*—the brain is undergoing an actual physical change. A change that makes it easier and easier for the synapse to share chemical links, which spark together. This allows thoughts to trigger faster and more easily. It is said that this is a microcosmic example of evolution, of how humankind and all sentient beings adapt and create adaptation.

The chemical that creates the sparks utilized to transport information is known as nerve glue. Nerve glue is a reusable commodity utilized by the brain. What is interesting in relation to the idea of abundance and this being an abundant world, is that this nerve glue is not in abundant supply—it is limited in quantity. Whatever amount the brain has, the brain has, so the brain must reuse its nerve glue over and over. Why is this scientific fact important to the establishment of spiritual principle, and ultimately, soul mastery? Why is it important to the development of a consciousness of abundance and conscious neutrality?

It is because, when you have an a-ha moment and make the choice to initiate a new pattern of thought and feeling, a change must occur in your brain's current neural patterning in order to establish the new thought. For example, when a new pattern of thought is initiated, a thought such as, "I am an abundant being living in an abundant universe."

What happens is that the brain begins to take the nerve glue from an old thought—I am poor—and starts using it in the creation of a new neural pathway that will support the new thought of abundance. The more the new thought of abundance becomes entrenched in the brain as a neural pathway, the less nerve glue there is available to hold the old thought, attached to poverty and lack, has to hold it in place.

When enough glue is removed, the old neural pathway and old belief pattern of lack, poverty, and limitation dissolves; for the new pattern of abundance is established.

However, for this to occur, you must make a conscious choice to reinforce what you are learning through practical application and repetition. It is not a one-time thought! It is the repetition of the new thought that reinforces the creation of the new belief. Repetition initiates firing and wiring and over time, this repetitive action creates new neural pathways and networks. And while doing so, the old, outdated thoughts of lack are being dissolved as the new thoughts of abundance are established.

"Thoughts reinforced over a period, reshape your brain, and thus change the physical construct of the landscape of your reality."[25] And because the body is engaged with the subconscious mind, the latest brain research is showing that you must feel and experience interiorly, in the now moment, to cause effect. To capture the highest possibility available, that feeling must be an elevated emotion—not one oriented toward fear, lack, limitation, and so forth—but an elevated emotion of gratitude, appreciation, trust, satisfaction, etc. You must feel that elevated emotion first.

Realize the experience interiorly in the now moment as happening now: feel yourself as loved-now; feel yourself as whole-now; and feel yourself empowered in a work you love-now; feel yourself as abundant-now. Doing this sets up an alignment with a higher field of potential.

As you practice this, through repetition, the brain and body are no longer a record of the past steeped in negative-oriented past perceptions, but they are anchored in the now—causing effect—an effect that will manifest a new reality, for you are by doing this, rewiring the brain in the now moment. And the brain doesn't know the difference between the past, now, or future events. It calls them all into the now moment, in the same way. Call the future into the now! Cause effect! Consciously choose to do the work in consciousness that is required.

Order is the first Law of the Universe...Vibration is a creative law...
The whole universe is in vibration, and that vibration is under law.
Chaos would result if the law were not supreme...
Each thing has its rate of vibration...
But what causes vibration? MIND! [26]

~

Your consciousness changes when the energy and vibration you carry shifts! As you begin to view things from a higher dimensional energetic, vibrational, and frequency perspective, you start to see or sense experiences, people, and situations in a new way.

---

[25] Steve Parton, *The Science of Happiness*, internet article
[26] Charles Fillmore, *Jesus Christ Heals*, Chapter 11

In the realization that came with the understanding that everything is energy, vibration, and frequency, I began to see with great clarity that I lived in a vibrational universe and that I was a vibrational being having a spiritual-vibrational experience in a human form. What I held as a vibration—in the form of thought, feelings, and emotions—mattered to my soul growth and awareness. It mattered to the creation and development of both my spiritual experience and my experience of physical reality.

Energy and vibration don't lie—they reveal! But they can only reveal their secrets and hidden wisdom to those who have eyes to see. Energy doesn't lie, also means, you can't say one thing and feel or act the opposite of what you just said. The true energy of the situation will be revealed, to those who are sensitive to energy!

Every interaction that occurs in your life is an exchange of vibrational energy and has the capacity to transform you. Your vibration attracts (or repels) people, experiences, your current landscape of reality, and more.

Your vibration affects how you show up, how you interact with people and situations, and why you do what you do! Your vibrational field holds the essence and consciousness of who you are. You are a vibrational being having a human experience! You are a vibrational being emitting energetic signals all the time, and energy doesn't lie!

When you consciously, begin to claim dominion of the I AM, through being willing to transformation "old" beliefs and perceptions, this choice has the capacity to spiritually and emotionally mature you. How? Whether you realized it, or not, you enter into an agreement of conscious surrender. You agree to the release the attachments that have been previously claimed by the 3rd dimensional, survival-oriented, adverse ego.

This act of agreement, with its choice to change, begins to initiate a new sense of identity, and belief agreement, as it shifts you from an unconscious agreement and engagement with collective third-dimensional energies, into a more conscious, entry-level fourth-dimensional energy and awareness (each dimension has multi-levels of awareness).

The dimensional level at which you claim the dominion of the I AM, as you (and you can claim it, by denying it), creates a corollary vibrational agreement, alignment, and momentum, that creates your world. Together, they act as "creation in action."

As an example, if your claim of dominion is in alignment with the third-dimensional, fear-filled, adverse ego, small self, you become subject to encountering those types of experiences that continually justify who the small self believes it is—not worthy, poor, not enough, ugly, superior, beautiful, perfect, big shot, poor me, pompous, victim, etc.

The adverse ego, small self takes great joy in gathering evidence through experiences that are in alignment with who it believes it is. The adverse ego then continually replicates these experiences in order to "feed" off the energetic evidence of those experiences. Thus, being able to feel right and justified, in its continued claim of fear-filled energies.

However, when your claim of dominion, is the I AM-Christ Consciousness; you claim a resonance with a higher vibrational and dimensional field of awareness. In higher levels of energy, instead of an alignment to any type of fear energy, you claim an alignment to the energies of appreciation, gratitude, peace, love, truth, authenticity, transparency, compassion, collaboration, etc., as who you are and how you show-up.

You call forth a demonstration that is in vibrational alignment with the wonder, awe, and possibility that supports the new energies. All done without any sense of, or need for, justification, pride, being seen, or being seen as right. As you stop agreeing to, or affirming, the untruths of third-dimensional fear energy, you begin to perceive a new field of vibratory energy arising around you, a new reality coming into existence. This new field of energy is a living reality that already exists, and that operates outside of the fear agreement. Your work is to learn how-to align with this field of energy and call it into being.

Soul mastery, and fifth-dimensional awareness, claims, holds, and is a vibrational agreement to, a whole new identity, and value system. A value system that knows that all levels of dimensions, all light, all awareness, all initiations, and all information required for soul mastery, spiritual growth, New Earth consciousness development, soul evolution, etc. are available: right here, right now, in this present moment, and that it can be no other way! This will be discussed in greater detail a little later on in the chapter.

Be impeccable with your words.
[and thoughts, feeling, emotions, behaviors, and actions.]
Don Miguel Ruiz

~

Additionally, as a vibrational being if you say, "I am love in expression," and then immediately get angry or feel anger toward someone, that creates a space of incoherent energy. And it will be felt. Or if two people are fighting and they assure someone who has just walked into the room that everything is just fine, regardless of the words, the true energy in the room will be felt by someone who is sensitive to vibrations.

This is not about right or wrong—it is about creating alignment with what you say and feel and with who you are. If you affirm, "I am abundance in action," then continually bemoan how in debt you are, this creates a resistance in the energetic field of your being and words against that which you are working to demonstrate. From an energetic perspective, that energy of resistance will set up a subtle vibrational signal that is incoherent and will vibrationally thwart the attraction of higher ideas and possibilities to you. And those with ears to hear and eyes to see will sense the resistant energy in your field regardless of the happy words or feelings you speak and share.

Someone will take notice of the energetic incoherence between your energy field and your words and actions. This is not about a one-time event—this is about showing up consistently in an incoherent way and yet believing that you are coherent. If you are "lucky" or "blessed," someone will point out the incoherency to you. At first, you probably won't like it. But as you spiritually awaken, you will appreciate the feedback and begin to see it more clearly for yourself. This act of seeing for yourself or "looking" is known from a spiritual context as honest self-observation and being your own witness.

Honest self-observation is a spiritual milestone because spiritual growth can only truly begin when you start to practice self-observation. It means honestly observing, without qualification or excuses, all your thoughts, feelings, behaviors, and actions. If you can see it, you don't have to be it! You don't have to be it if you can see it, for in that moment you can make a new choice as to what you energetically want to create and demonstrate in your life experiences. And that is where real spiritual and emotional growth and maturity occurs.

Over the decades, I have come to realize that it is not enough to see your own incoherent, unskilled behaviors or actions and then apologize for them happening. Lots of people see their incoherent (unruly, unkind, bad, negative) behaviors and apologize repeatedly. "I promise I will never hit you again." Or they continually answer their phones and text when they have been asked to shut them off or put them on vibrate. They apologize, say they are sorry for their actions, but they do not take any action to change their behaviors. It is only when you honestly observe the incoherence in your thoughts, feelings, or actions—and then make a conscious choice to take an action to support a shift of your energy through your behaviors and actions—that a true change can occur. It is only when you choose to shift into coherence, through the alignment of the energy of your thoughts and feelings with your words, and actions, that you can begin to experience alchemical transformation.

If I had one hour to solve a problem, I would spend 55 minutes developing the right question, for when you have the right question, the answer comes easily!

~

As you begin to view things from an energetic, vibrational, and frequency perspective, you start to ask questions that are more in alignment with that new context. Questions arise such as: What might energetically be happening here? What interesting dynamic might be arising here? What meaning am I giving this? Which is quite different from: Who did this? Who is to blame? Why, why did this happen to me?

Shifting to this type of question invites a subtle shift in perspective and meaning in the moment. I call these types of questions, empowering questions. Empowering questions have the capacity to shift perception and move your mind-focus from a third-dimensional context of—What just happened here? Who is right or wrong? Who is to blame? Or, how can I be right in this situation? —to perceiving the situation from a sense of non-attachment, non-resistance, and not taking anything personally in the moment.

An empowering question has the capacity to open the energetic field and invite a new sense (experience) and perception. You perceive things differently, because an empowering question allows a wisdom to arise that comes from a higher perspective. That perspective is both spiritual and scientific in orientation. It is scientific in the sense that it has the capacity to stop the amygdala's fight-or-flight response and to kickstart a response from the executive centers in your frontal lobe. And it is spiritual in the sense that shifting to the executive response center opens the field to a higher-wisdom answer. Why? Because the field of resistance has been decreased or eliminated. Empowering questions are questions such as:

What is important here? What is the possibility here?
What is more interesting—being right, or being a contribution?
What meaning am I giving to this?
What might be energetically happening here?
How might I be a contribution?
Is what I am choosing a contribution for the good of all?
How might I be a support?
This is interesting, what might be happening, energetically?
What would a person who is ___ (abundant, centered, excited about life, etc.) be feeling right now?

Remember, everything is energy, vibration, and frequency. You live in a vibrational universe and you are a vibrational being having a spiritual-vibrational experience of evolution in a human form. Know that soul evolution is as much about your physical form (what you do and the actions you take) as it is about your non-physical form (the energy and vibration you take your actions from). It is said that the dash between birth and death (19__ – 20__) is a short span, use it wisely!

Things are manifested in your life in alignment with the vibrational energy (thought and feeling) that you hold regarding who you claim to be. Do you claim, I am worthy to receive, or not? It is your choice, for the law works. Right use or reverse use, the law works!

When you claim something as awful, you receive the residual effect of that awfulness. You are in [the law of] exchange.
When you claim something as holy,
the holiness that is implicit in what you see blesses you back.[27]

~

The ability to read energy is neither strange nor unconventional, nor it is something special that only a few people can attain. Reading energy is an inherent yet latent talent and the more we awaken to higher levels of awareness, the more awakened that talent becomes. Along with other talents, too. Have you ever thought of someone and the phone rings, and you pick it up only to find out it's them? That is telepathy at work in you!

These are natural responses of being when you are in energetic alignment and connection to Universal Source Energy. The more you are willing to dissolve and release third-dimensional baggage and to

---

[27] Paul Selig, *The Book of Mastery,* 3 of 3 of Master series

release resistance, the less energetic static there is to block or interfere with higher dimensional messages and intuitive hints that are always being broadcasted. Remember, you stand in the midst of all things now! Higher potentials and possibilities are available to you now! As you consciously remove the energetic blocks, you will free up space for new ideas to flow through you. When you are an open and clear channel, energy flows.

Everything is energy, vibration, and frequency. It is how the universe is set up and how it works. Stated very simply, from a metaphysical perspective, energy is the dynamic substance that underlies, interpenetrates, and ultimately, interconnects all things; and energy is set into motion through vibration.

Vibration is a rate of motion and that motion creates a momentum, or movement, of energy. Vibrational momentum holds a speed, slow/dense or high/light, and the vibrational momentum speed aligns with and emits a frequency. Vibrational momentum akin to a radio tuned to a station's specific frequency. You can tune into all different types of stations, and that is done by turning the dial to tune into their specific frequencies. Might this be the same with consciousness? Change the frequency, change your programming!

The level of frequency determines the amount of knowledge, light, and wisdom contained within the vibration and its frequency range. The frequency range determines the dimensional level at which the energy is vibrating. And energy doesn't lie.

Know that the shifting dimensional perception occurs through consciousness. How you perceive these dimensions will determine what information (light/wisdom/knowledge) you allow to come into you and through you. Remember, it is the vibrational intention and vibrational atmosphere that you allow to be present through your consciousness that underpins what you do and that makes the difference in working with energy and vibration.

> *I believe that what Jesus and Mohammed and Buddha*
> *and all the rest said was right.*
> *It's just that the translations have gone wrong.*
> John Lennon

~

You cannot say, "I know I vibrate at a fourth-dimensional level," and then have all your words, emotions, and actions be third-dimensional lower survival-oriented expressions. People will notice the incoherence. This means that you cannot say you stand for love and act out hateful actions or walk around with a chip on your shoulder, reacting to every little thing that is said that you don't agree with. People will feel the incoherence that is emitted through your energy field.

Your energy field is an energetic field that surrounds your body and that is palpable to those who sense energy. It was often said of the Master Teacher Jesus, ". . . and he read their hearts." Which, in today's language, means that he read their field of energy. Then he knew what to do in any given situation—and often he just left.

> *You cannot solve a problem*
> *on the same level of consciousness that created it.*
> Albert Einstein
>
> ~

Charles Fillmore shared nearly a hundred years ago that "*Unity [metaphysical teachings] is a bridge between the third and fifth dimension.*" Meaning, metaphysical interpretation of life, and its experiences, is a fourth-dimensional process that initiates the call into higher dimensional awareness. I found this to be an extremely interesting concept at the time, and yet I did not quite understand the impact it would have on my life.

Years later, I discovered that both Mr. Fillmore and Ernest Holmes (founder of Centers for Spiritual Living) shared in their writings the assertion that dimensions are not locations or places, but levels of consciousness that vibrate at different rates.

This statement, along with a few other events, supported an unlocking of mind-memories that had previously been shut away, under lock and key, in the deep recesses of my consciousness. And an expansiveness of consciousness began to take place. I discovered that lower fourth dimension is a more metaphysical-like energy, and the higher you go in multi-dimensional awareness, the more mystical and instantaneous the energy becomes. Mystical visions and synchro-divine events occurred (those unexpected moments when the Universal Intelligence aligns people, situations, and conditions and the seemingly miraculous occurs and it transforms you).

Simultaneous realities appeared, revealing the different outcomes available based on my choices, as did a desire for transparency-authenticity in all my undertakings, telepathy, and other miracles beyond my normal comprehension. All of this started to be normal, not special. I began to explore new dimensions of understanding regarding the life of the Master Teacher Jesus. I discovered that whether I believed Jesus lived or did not live was of no consequence. What was important was that I saw the spiritual legacy that was left through the mystical writings and stories regarding his life experiences and that there was profound teaching hidden within them.

As my eyes and heart opened, I began to see that this legacy was a blueprint, Christ template for soul evolution, soul mastery, and the journey into higher realms and dimensions of awareness—the Kingdom of Heavens — metaphysic interpretation was the bridge across the dimensions.

Dimensions are not places or locations
but levels of consciousness, each vibrating at different rates. [28] & [29]

~

The following chart, Example A, depicts that dynamic movement of energy, vibration, and frequency across dimensions. Each dimension is held in place by the energy, light, information, and knowledge that match the underlying energy, vibration, and frequencies. These work together to create and hold the foundational thought and belief constructs held within the dimensions—third, fourth, fifth, etc. (Refer to Example B for a list of those constructs and beliefs.)

Example A shows, in a simple format per Fillmore, where the metaphysical movement sits within the dimensions as a construct. Metaphysics and its inherent teachings are said to be a bridge that has the capacity to lift you out of third-dimensional thinking and move you into the multiple levels and portals held within fourth-dimensional awareness. As a person progresses (like the Master Teachers Jesus and Buddha), they move into higher and higher expressions of fourth-dimensional and fifth-dimensional energy fields.

---

[28] Charles Fillmore, *4th Dimension in Man*, Unity, Unpublished, Unity Archives
[29] Ernest Holmes, *Science of Mind*, Science of Mind Publishing, Definitions-Vibration

<u>Example A: A Universal Multi-dimensional* Matrix**</u>

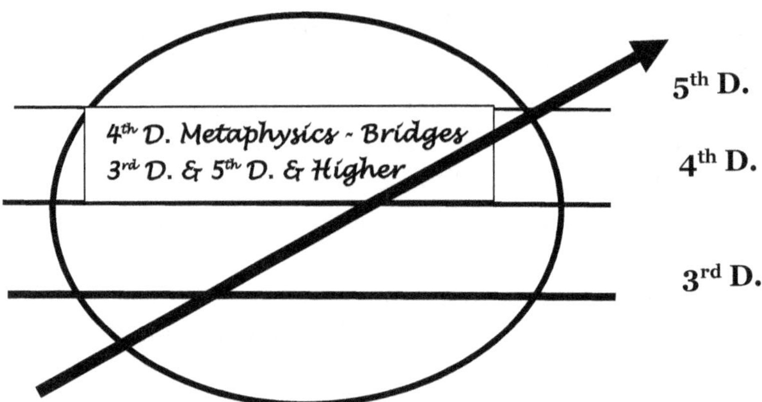

*Dimensions are not locations, they are vibrating fields of energy, with multiple frequency ranges contained within them.

**Matrix: an environment in which something develops.

The higher the dimension, the finer the frequency and energy that is in that dimension (as represented by the lightening of the representative dimensional lines). One who tunes into that frequency field or range sees from a state of awareness that matches the information and knowledge that the frequency range holds. Alignment with higher dimensional awareness speeds up energy and has the capacity to manifest things more quickly. At higher frequencies, energy moves faster.

As each dimension holds different frequencies, each requires different capacities to sustain the information, knowledge, and wisdom contained within its range. The work involved in the creation of mastery in consciousness over the elements of the third dimension is not about focusing on how to solve the problems of your life. The work is to master learning how to shift the energy, vibration, and frequency from which you view and participate with life. Once you demonstrate or master a higher level of understanding in your consciousness, it becomes a part of your energy field or range, and it lives in your cellular memories.

Thus, it acts vibrationally as a point of attraction for the working of the law on your behalf. The law then goes forth into the greater field to match up the cooperative components required for the manifestation of your request, thereby matching up and drawing to you all the components (people, situations, conditions, etc.) that are in alignment with your vibrational signal. The spiritual path, the spiritual journey, the journey of awakening, the path to ascension—whatever name you choose to use—is a path of rising frequencies. It is a path of ascension in consciousness and awareness. It is about attaining new levels of frequency, entering new dimensions of understanding, and opening new portals of information.

However, this energetic rising in consciousness requires that we master each level of frequency to sustain it and hold it steady in our mind. Once a new level of awareness is mastered, it then works effortlessly in our life. It is not unusual to have what appears to be an instantaneous awakening moment. Although it happens in an instant, know that there have been years of conscious and unconscious preparation behind it. If this happens to you as an individual, then what does it mean for the collective whole as a species?

As a species, there is a collective awakening occurring. People are shifting into new levels of lower fourth-dimensional awareness. Many are observing that things are speeding up and manifesting faster, and they are sensing changes within them. What does an energetic movement into higher dimensions feel like? Here a few of the signs that people are sharing.

Do you sense that you are strangers in a strange land? Do you find things that have always worked are no longer working? Do you notice that you are no longer relating to people you have always related to? Or do things you used to enjoy no longer have the verve they once did? Do you often find that rather than wanting to run around thither and yon, you find yourself wanting to stay home and BE? Have you noticed or had a sense that your energy is shifting?

Is your body having strange sensations, like being tired even when you know you are not? Do you feel scattered when you thought you were clear? Do thoughts disappear out of your mind and, like whoosh, they are gone? Do you feel anxious when you know you aren't? Do you feel sad when you know you aren't? Do not despair. You are going through some energetic dimensional shifts. Be appreciative that you are sensitive enough to be aware that something is going on.

Learn how to see how everything connects to everything else.
Leonardo Da Vinci

~

Another sign of awakening that occurs, is that of noticing, when you look at a clock or a group of numbers, that the numbers are sequential. 5:55, 1:11, 3:33! These are often "hidden" messages affirming your connection to Source Energy. The following is my personal interpretation of what it means when sequential numbers appear:

1:11 = angels (good thoughts) are here;
2:22 = good or a sense of balance desires to express, engage it;
3:33 = masters and guides are nearby;
4:44 = foundation, stability, more good thoughts/angels;
5:55 = change is in the wind;   6:66 = balance and observing;
7:77 = miracles and success coming;
8:88 = progress, moving forward; 9:99 = complete on some level;
10:10 =one with Source—energy is increasing exponentially;
11:11 =mastery—your Inner Guide is here, ask a question;
12:12 =mastery and spiritual completion.

If you are experiencing these signs, know that you are open to or opening to higher dimensions of awareness. Your work is to consciously chose in each moment to keep your vibration focused on elevated thoughts/feelings. Do not lower your energy by engaging in the petty exchanges of everyday life experiences. Honestly, observe when your desire is to be seen as right or special and remain above the fray of chaotic discord that tries to seduce you into its clutches. You need no protection except the power of the open heart, a power that arises from your connection with universal higher-heart intelligence.

Think of God as a Cosmic Mind....
a great Universal Principle back of this universe....
Jesus when he came in contact with it, called it.... the Father. [30]

~

Astrologers and other metaphysical teachers are saying that since the eclipse of 2012, there are multiple types of evolutionary shifts going on. Earth shifts, body shifts, DNA shifts, climate shifts, seismic shifts, dimensional shifts, perceptional shifts, brain-wiring shifts, and sleep-state shifts, to name a few. These shifts are occurring more and more rapidly.

Fillmore also said that the subconscious mind was designed to take imprints from Source Energy, but because of freewill you have chosen to take direction from third-dimensional experiences and awareness. And that choice has created the discord and inharmonic resonance that you and many others are experiencing. It is time to *"lift up your eyes and see fields ripe for harvest."*

---

[30] Charles Fillmore, *Sunday Talk,* January 20, 1920

It is time to see that being open to new levels of multidimensional experiences and thinking will move you in new directions that will evolve you. It appears that the upward spiral of evolution is occurring now. The changes being experienced are shifts into higher dimensional awareness, and they portend the new information that is being vibrationally imprinted regarding who we are. These shifts are occurring globally, collectively, and individually. They are inviting you to let go and to learn how to dance at the edge of mystery. However, even though many are awakening, there are still many who do not have the tools to recognize what is happening nor the skills to manage it.

As recently as twenty years ago, if someone would have spoken to me about dimensional shifts and multidimensional aspects of being, I might have laughed and said, "Oh, a New Age hippie expounds again."

After having my own experiences with multidimensional shifting, I am now an empirical knower. Aspects of my being have been opened, gifts have been awakened, imprints have been downloaded, and mysteries have been unlocked. Once the proverbial cat is out of the bag, it is difficult to get it back in—and frankly, I do not want to go back.

This is an evolutionary time. You are in an evolutionary cycle, and this cycle will not allow status quo to remain. Therefore, the work is not about looking for what is right or wrong or who is right or wrong. It is about conscious evolution. How are the experiences in front of me serving as my teachers? How is the vibration I am currently experiencing serving my soul growth? Does this vibration, feeling, or emotion serve greater alignment?

You are energy, an energetic being, an electromagnetic field of walking-talking vibrational energy. The electromagnetic field around you is specific to you and contains vibrationally encoded information and data and, whether you see it or not, it is there radiating around and from you. As a field of energy, you are broadcasting signals all the time.

As an energy field, you attract what you place your attention and intention on, and this makes all the difference when it comes to being and living as one who attracts greater good, abundance, and love—or not. The work is to learn to manage your vibration and your field of energy—and to be more sensitive to the presence of resistance when it shows up in your life experiences. Resistance slows energy movement.

When you say you want to grow and change and then do things that are out of alignment with who you say you are or what you want to experience, you will feel the kickback of the two energies colliding (the energies of wanting to flow with things yet being in resistance to them). Be sensitive to it. Do not ignore it. Sense it and be interested in the possibility that you might be out of alignment.

The learning that comes as you traverse the landscapes of new realities is that you cannot take your third-dimensional baggage with you. You must be willing to mature and release the limitations of your will/ego to move forward. To continue the journey into higher dimensional awareness, you must be conscious, awake, aware, and cognizant of how you are showing up in each moment. To do that requires that you live with a sense of clarity, intentionality, coherency, transparency, authenticity, and accountability in everything you do.

Remember, all dimensions exist right here, right now! That is why there is nothing to get, only layers of muck to peel away. That unpeeling then reveals a new level of light, understanding, and awareness—which shifts how you see your current reality. This is not about comparing right or wrong. There is no right or wrong; the reality you choose to live from is simply where you reside until you don't. The work, as you traverse the spiritual journey, is to awaken you to new dimensions of reality. Interestingly, most people do not choose to awake until they are in the midst of some great emotional drama. This can be a level of resistance with no apparent solution at hand, a traumatic situation, a dark night of the soul, a dark night of the ego, or what have you. Each of these experiences has the inherent capacity to call forth a yearning to open the heart to a new level of awareness for life and its experiences.

Physical Body, Light-Body, and Dimensions

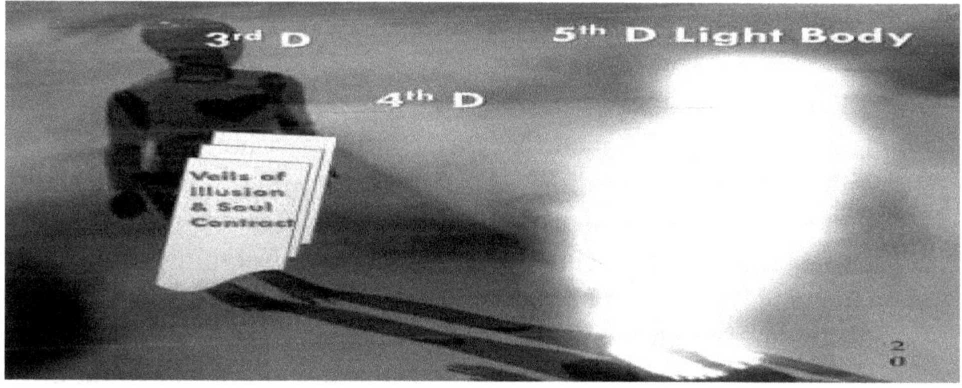

Third Dimension/3-D (Realm of Appearance):

Requires the transmutation and dissolving of the dense energy of perceived separation and the embedded perceptual beliefs of humankind and family dynamics, which includes all attachment to "My Personal Story." You are here to pierce the veils of illusion and to learn how to ascend into higher dimensions of awareness & be a master facilitator of energy.

Fourth Dimension/4/D (Activation and Integration through Intention):

Activates inner soul abilities and aligns with higher dimensional constructs and knowing. This dimension has multiple states and levels within its realm of awareness, including lower fourth, mid fourth, higher fourth, and more. A pivotal realm, it can feel like walking between two worlds until fourth-dimensional energies are anchored and embodied cellularly.

Fifth Dimension/5-D (Christ, Light Body):

This holds the blueprint and template of wholeness and expression of "perfection." No-thing consciousness: no story, no fear, no-thing to block light energy. There are multiple dimensions beyond this one and all are supporting this movement into the next phase of conscious evolution of humankind. Holding this consciousness does not mean you will not have a body.

The following overview of third-, fourth-, and fifth-dimensional influences, behaviors, and actions are intended to invite you into observing the differences between the dimensions and honestly considering what shifts might be required for you to engage a new level of awareness!

3-D

If you always do what you've always done,
you will always get what you've always gotten.
Unknown

Currently, much of the collective whole of society lives influenced by a third-dimensional view of physical reality. In the third dimension, energy is dense and heavy. It is often difficult to work with and it takes significant effort to make changes. Yet things are constantly changing. It is a human, survival, ego-oriented dimension, steeped in a linear and Newtonian-oriented awareness. It is a dimension where many walk around unconscious and unaware of the glory and mystery that lives within them.

People enmeshed in third-dimensional thinking choose to believe that what is happening is *real* and, thus, judgments are made as to what is right and what is wrong. Great meaning is then given to these judgments. It is a reality where people say, "I can't, you have to do this for me." "I am not responsible for what just happened." Thus, the dynamics of a lack of responsibility and accountability for actions abounds.

Additionally, it is a realm of poor me, me-me-me, what I want, being a victim of life and societal and cultural impacting. Yet the adverse ego/small self wants to be in control and, desperately at times, works to be in control through conscious manipulation of events and people. The ego does this in order to be in control, special, and right. Even if being in control, being special, and being right makes them appear as a wounded victim of circumstance.

Third-dimensional expressions are full of resistance and are, collectively, in the midst a resistance epidemic. The resistance is expressed as being wounded, needing to blame others, wanting to project fault on others, and at times even feeling the right to emotionally vomit on others.

People steeped in third-dimensional awareness predominantly see through a fear-based lens and view things in accordance with the prescribed perceptions they have created. Fear is a general label used to describe any dense energy that holds you hostage to lower 3rd dimensional vibrations (fear, anger, anxiety, guilt, greed, projection, shame, blame, jealousy, judging, manipulation, control, righteousness, overwhelming need to be right-safe-seen, addictive emotions, etc.). Fear activates in order to show you where your belief and perceptual distortions are. The question is, are you awake enough to see that is what is happening?

Fear-based encounters lower your vibrations and create dense energy. The dimensional level that you dominantly resonant with, is in direction correlation to the amount of dense energy your body holds. Perceptions and beliefs hold great meaning—and as they are held and relived in the mind—they become more deeply embedded as a structure of emotional knowing.

Your degree of reaction, and the subsequent action you take after the reaction, is a signal of how much fear, and dense energy, is embedded as a belief, perception, or programming. To move to higher levels of awareness, you must make the conscious choice to release and dissolve the dense energy; for dense energy blocks light energy movement and suppresses next-level soul growth. And the more light and higher frequencies you embody, the faster you progress.

Third-dimension reality is where people believe they need to know. Know answers, know they are safe, know they are right, and so on. When you believe that there is something that you need to know, that you have a *right* to know, what happens to your body? Your muscles tighten, your emotional reactions often go off the chart, and you become determined to be in control of a person or a situation. You become self-righteous and self-justified. This need to know creates an emotional vibrational resistance within the body system. At every dimensional level you are given a vibrational frequency choice. *Do I choose to make a decision from my currently level of energy—and remain in the status quo—or do I want to make a choice from a higher level and get different results?* You use that choice often and it is always based on the dominant vibration held within your being at any given moment.

Third-dimensional alignment is steeped in the energies that align with fear, sadness, anger, greed, and jealously. These are not wrong or bad emotions; they are simply energy vibrating at a lower frequency. All emotions were initially designed to be influencers and directors of your experiences for making higher choices, but over time you lost this information and began to give these lower survival emotions great meaning based on your personal judgment of what they mean in the moment. You began to see them as something against me, rather than as something that is actually for me—for good.

The third dimension has embedded structures of knowing and rules as a part of its frequency level. Embedded third-dimensional energetic information is rooted in competition, lack, fear, guilt, blame, defensiveness, me-me-me, resistance, linear time, challenge, limitation, and other fear-based limiting perceptions and structures of knowing. Additionally, these perceptions live not only in you as an individual, but also in the collective consciousness.

The third dimension holds programmed patterns of meaning regarding relationships: meet someone, fall in love, believe this is the only one. Then comes the controlling, the cheating, etc., all of which are steeped in a construct of conditioned love. Looking for love in all the wrong the places holds this energy in place.

The third dimension firmly holds the constructs of duality—good/bad, right/wrong, judgment based on past perceptions, etc. Thus, you create a future that is determined by these judgments and perceptions. It is a reality steeped in survival and, over time, it has wired your brain for survival as the primary reaction to the experiences of life. This survival mechanism teaches you to look for threats even when there aren't any. Because of this, reaction and resistance are your primary responses. All of this exists because fear rules as the dominant energy in this dimension. The belief in your capacity for choice and self-responsibility are negligible and often overridden by deeply embedded thoughts of fear.

The third dimension holds the embedded patterns of the adverse ego, a.k.a. your small self, the adversary, even the devil. The adverse ego plays small (or extra-large as a bully or power-monger) for it believes it must in be charge of the world to get what it deserves, or that the world or other people will take advantage of them in some way. Our adverse ego/small self rules according to beliefs and perceptions that have been passed down through generations of both personal family and collective society. These beliefs include ideas regarding culture, country, gender, race, and thought-feelings such as *I am not enough, I am not good enough, I do not count, I will always be poor because my family is poor*, and so forth.

The third dimension supports advocacy for change that is steeped in fear, anger, separation of systems and ideals, and force. This includes the desire to change things from a perspective of force that includes threats, war, forcing change, marching in anger, ugly signs depicting words and graphics meant to hurt a group or person, and so much more.

The perception of third-dimensional reality is impacted by all the aforementioned until you are ready to make a conscious choice toward change. This change occurs when you are ready to shift from a fear-based reality to a joy-oriented, elevated-emotion reality. Elevated emotions are inspired by Source Energy and thus, arise from a higher level of awareness, coherence, and harmonic resonance.

Elevated emotions, also holds higher level qualities, and values. This includes trust, joy, peace, appreciation, gratitude, compassion, inspiration, intuition, clarity, integrity, intentionality, collaboration, cooperation, and more. To hold elevated emotions means that you have mastered or overcome human-ego-survival-oriented emotional reactions and have evolved your capacity to respond.

Elevated emotions hold a fourth-dimensional vibration that is more heart-centered, thus they are more coherent, purer, and unconditioned than the lower third-dimension-oriented survival emotions embedded below the heart center.

Example A: Elevated Emotions (4th D.) vs. Lower Emotions (3rd D.)

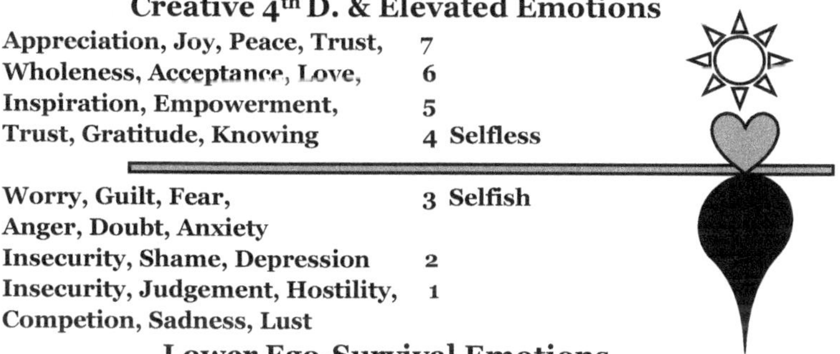

Brain research reveals (refer to work by HeartMath, Dr. Mark Waldman, or Dr. Joe Dispenza) that the brain doesn't know the difference between past reality, current reality, or future reality—it is all experienced as *now*. If every day you relive yesterday's negative events (or the last decade's), you bring all those emotions into the now. And you will be creating vibrational broadcast signals filled with vibrational data that reinforces old habits and negative emotions you hold around them. Then you wonder why life doesn't change for you. It can't change if you don't change it!

4-D

*The Invisible Realm is real and tangible
to my mind and feelings, now.* [31]

The fourth dimension is the initiation into the realm of metaphysics, the realm beyond the physical; it is the bridge between the third and fifth dimensions. Here, you begin to dance with innovative ideas and ways of being and yet you are still dancing between two worlds.

The fourth dimension holds multiple levels of awareness. There are realms such as the lower fourth, mid fourth, and higher fourth. It is in the fourth dimension that you consciously begin the journey into soul mastery and that the full collapsing of third-dimensional energies, beliefs, and perceptions starts to happen. However, the initial journey begins in the third as you have an awakening or say-YES experience. Here are three of the most common ways that the collapse of the third-dimensional energies occurs:

1. Circumstances Force Collapse: as the spiritual journey into soul mastery is initiated, the initial awakening begins to cause things to collapse around you, and it appears as if circumstances are forcing you to make a change or to shift—this occurrence may be a divorce, illness, death of a loved one, loss of job, etc. It is a choice point moment: if you don't shift, the suffering of the third dimension continues.

2. Conscious Choice: as you grow in consciousness and go a little further into fourth-dimensional awareness, you understand that you have the power to consciously shift situations and experiences. So, you consciously make a choice to do so, or not, while holding the highest and best for all involved. And how much you can affect a situation depends upon the level of fear, negative beliefs, etc. that you still hold in consciousness.

3. Intentionality: as you shift into higher levels of dimensional knowing and understanding, your focus is intentional and holds clarity as to the role of energy, vibration, and frequency in demonstrating and shifting situations in your current reality and, thus, you consciously participate in alchemical transformation and it is an instantaneous experience.

The fourth dimension is akin to a bridge that you must cross over to enter the fifth dimension. However, you do not just blatantly walk

---

[31] Charles Fillmore, *4th Dimension in Man*, Unity Publishing

across the bridge. You must engage in conscious spiritual growth and awareness on multiple levels of being in order to initiate a movement into fifth-dimensional awareness. Many enter the fourth but get stuck in the entry-level energetic requirements and this halts their progress (for a time or for a lifetime). At the higher-fourth level, you begin to tap more consistently into fifth-dimensional reality.

The fourth dimension holds lighter energy than the third and it is in the lower fourth-dimensional realms that the initiations of release begin. You will begin the release of the old baggage of negative perception, shadows, lower thoughts and feelings—all of which are energy that will be dissolved to be lifted into higher-level vibration. Many at this time of breaking-down of old mental and emotional structures may experience, phenomena. Phenomena is seeing colors, auras, lights, visions, and more. Do not get caught up in them, they are part of the process, for some. They do not mean —in any way, shape, or form—that you are special, chosen, etc. It means you are shifting consciousness, that is all! Many do it without phenomena.

The lighter fourth dimensional energy starts to draw things more easily and effortlessly to you. You will process things faster, too! But do not let that fool you. This realm requires constant vigilance or you will slip back into third-dimensional thinking and ways of being, and you will wonder what just happened. Often it feels as if you are living or dancing between two worlds. And you are!

The fourth dimension is the realm of paradox, polarity-holding, and seeing things from a variety of different perspectives. As a bridge between two other dimensions, the fourth dimension bridges two different points of awareness or references and you willingly dance with polarities such as knowing/not knowing, wanting/letting go, dark/light, shadow/light, true/not true, sure/not sure, clear/murky, yes/no, and even outdated ideas regarding what masculine/feminine[32] means. You are now a holder or container for the idea of "both/and." As the idea of being a polarity holder comes alive in you, you will notice that you naturally tend to look for multiple creative solutions for issues and that you have released your attachment to the third-dimensional construct of either/or. You know that you are making progress in crossing the bridge when you realize that you do not have to make someone else wrong in order to feel right about things.

---

[32] *Embracing the Feminine Nature of the Divine* Toni G. Boehm. Inner Visioning Press. Amazon.com

The fourth dimension is the realm where you are invited to dance at the edge mystery and discover *what is*. As you are unwrapping the meaning of *what is alive in the moment,* you are also learning to energetically perceive and see what isn't interesting as a choice for your life!

Fourth-dimensional awareness is a place of questions, not answers. It is about remembering! You are reminded that what is energetically present in a space can change in a nanosecond. The fourth dimension invites you into seeing all your experiences without judgment, resistance, or attachment. This is the place of unraveling and disassembling old perceptions, beliefs, and attitudes. This disassembling of the old structures of knowing is an imperative of this dimension and must occur if you want to keep your energy light—so continually make conscious choices to stay in vibrational alignment.

Fourth-dimensional awareness initiates a capacity to see the hidden mysteries contained within various religions and spiritual teachings. You will observe them in a new light and context, regardless of whether they are from Buddha, the Master Teacher Jesus, Christianity, Judaism, New Thought, or other venues. As you discover their hidden meaning, you realize the interconnectedness of all spiritual truth and principles.

Fourth-dimensional energy is lighter because you are letting go of old structures of knowing, which frees up energy for new gifts of grace to arise in you that you can use in other areas of your life. These gifts may include intuition, telepathy, being a master facilitator of energy, reading energy fields, seeing auras, third-eye awakening, seeing with more clarity across the dimensions into the past and future, interaction with inter-dimensional guides and master teachers, and more. As your gifts become more pronounced, they become a constant in your decision-making process. And although your decision and choices must be carried out in the third dimension, they produce outcomes that are better than expected because they have been made from a higher level of awareness.

Shifting into entry level fourth-dimensional awareness initiates another aspect of the journey. It is an alchemical transformation process, which entails, over time, a "deep" cellular memory cleansing, identity crisis, shadow exposure, and DNA restructuring. Alchemical transformation occurs around us and within us, all the time, but few know what it is or what it means.

Alchemy is simply the transformation of a substance, changing it from one form to another. Remember, the ancient alchemists who claimed to turn lead into gold? It was a hidden message and metaphor regarding the soul's journey. Alchemy occurs when the energy level or frequency that a substance is vibrating at changes—think ice to water or water to steam. Alchemical transformation is not woo-woo; it is a spiritual process.

During the alchemical transformation process, the mind and body are systemically prepared and transformed on all levels of awareness—physical, mental, emotional, and spiritual—in order to sustain and maintain higher levels of frequencies. At the same time, the neural pathways of the brain are rewired, reconfigured, and realigned to support new frequencies and information.

Initiation into higher fourth-dimensional and entry-level fifth-dimensional Soul mastery awareness, initiates an unwavering claim of dominion, that sustains and maintains the vibrational frequency of the I AM/Christ/Higher Awareness. This level of conscious awareness holds a steadfast knowing regarding the authority and dominion of who and what you are as the I AM, along with a full surrender to what it "takes" to be that I AM in action.

Fourth-dimensional energy often feels a little like insanity—and it is—for you are in transition, with one foot still in each world. Your ego is breaking down, you are releasing attachments to many things, you are detoxing on many levels as your light body is being activated, you are consciously engaging in seeking out distortions in your emotional field, taking risks of authenticity, you are letting go of the need to be safe and having others assure you that you are safe, you are rewriting your history and that feels discombobulating, and you want to sleep more as your body aches.

You know that your current physical reality is a response to your past and what you hold in consciousness. You start "challenging" karma, for you know it doesn't always have to this way. Multidimensional talents are awakening. You are honoring yourself as the I AM takes dominion of your awareness. The masculine aspects of being surrender to the more feminine[33] qualities of being (will, power, etc. surrender to trust, collaboration, cooperation, etc.). Synchro-divinity and synchro-

---

[33] *Embracing the Feminine Nature of the Divine*. Toni G. Boehm. Inner Visioning Press. Amazon.com

divine events (third-dimensional synchronicities) come into your awareness and become a constant companion.

You begin to process faster and the law of attraction/manifestation comes faster. Abundance starts to flow, and you realize that you are beginning to understand how to connect with the demonstration of it in mind—even though it may still be an initial stage of awareness. You now have the capacity to hold higher-level awareness and elevated emotions for longer periods of times. You notice that you are holding them until you are clear that your vibration is shifting, a new momentum is gathering, and an expansion of consciousness is occurring. Because of all of this, you are willing to do what it takes to clear and live through the initial expression of what feels like insanity.

The fourth dimension consciously awakens the innate aspects and qualities of being that are often thought of as leaning toward the feminine. Feminine aspects can include generosity of spirit, connectivity, interdependence, compassion, connection with the higher emotional intelligence of heart wisdom, the capacity to be conscious of when you are unconscious, the capacity to dance at the edge of mystery, to sit in the space of the unknown, and to be comfortable sitting with questions and to not need answers for now, for you are allowing answers to arise from the collective wisdom at hand. You are discovering the capacity to be conscious neutrality in action, which holds the capacity to neutralize negative energies in your presence—both your own and others. In this neutral and neutralizing consciousness, your need for advocacy shifts from radical anger and force for change to an understanding of the adage: *You cannot solve a problem with the same level of consciousness that created it!* Thus, you are more creative and peace-oriented in your solution development.

The fourth dimension is a realm of conscious (and unconscious) clearing and cleansing as you let go of resistance, fight, fear, and more. It is an activation reality, as the first movement into this realm of awareness activates the initial entry of "light energy" into the cellular structures of the body: energy that has been held at bay or hostage, by our fear-based perceptions. As it gathers momentum, this movement of light energy ultimately initiates a massive cellular shift and internal transformation within your being.

In fourth-dimensional awareness, you know that all power is in the now, and the more aligned you are with Source Energy, and the higher the frequency you consciously hold in mind—the faster the past dissolves! This dimension of awareness knows that dissolving or releasing past perceptions, beliefs, and behaviors can be as easy or difficult as you make it, depending upon the level of resistance you choose to engage and hold. Interestingly, there is a lot of talk about awakening, yet the soul doesn't actually "wake up." What feels or looks like awakening is the adverse ego's hold dissolving and a greater amount of light/ understanding/ knowledge/information/higher awareness coming into conscious awareness.

Once you tune into and align with higher dimensional awareness the soul, being as conscious as possible becomes a priority. You begin to recognize that you and others are either conscious or unconscious (not judging, just being aware). You notice that being unconscious is disempowering and supports disconnections, resistance, fear (fear is a belief that you stoke like a fire), and more. Fear and resistance disallow the free-flow of life energy and that is the reason the illusion of death appears in the third-dimensional constructs as a thing to fear.

The fourth-dimensional realm is home to the entrance into the unknown and it is where you will begin to feel comfortable with that journey. As you dance at the edge of mystery and dive deep into the interiors of being, dark nights of the ego, dark nights of the soul, shadow exploration, and other opportunities for clearing will occur. (Please note that the word *shadow* is more aligned as a third-dimensional term, for it has much baggage of interpretation associated with it.) These will occur in some form as you are working to activate and integrate the light energy of higher dimensional awareness into the cellular structures and memories of your being. During this phase, it is important to be aware that the process is only as difficult and uncomfortable as you make it through resistance.

The fourth dimension is a new level of feeling reality. It is not like the third-dimensional projection or owning feelings of anger, jealousy, safety, and so on. In the fourth dimension, it is a feeling of reality where you understand the imperative role that energy, vibration, and frequency (and thought/feeling energy as broadcast signals) have in attracting higher possibility and soul evolution.

The fourth dimension initiates a shift in consciousness that will ultimately move you from awakening to alignment to ascension, and as your understanding of reality shifts at each new level of awareness, you change! Know that this understanding of the movement between dimensions and levels is not a third-dimensional construct of comparison, of comparing how you rank against others; it is a descriptor and a tool for you to utilize in consciously assessing where you stand in each moment.

The fourth dimension initiates a new level of heart opening and feeling/emotional expansion. As your heart wisdom expands, you begin to emit a new vibrational signal, an electromagnetic field that supports alchemical transformation and higher vibrational matching of possibility in the unified field of consciousness. An open, expanded heart also holds elevated emotions, such as appreciation for all experiences, a trust that all is well, that life is just as it should be for the lessons you require, a joy for life, and more. These elevated emotions, each time they arise, release over 1,200 neuro-chemicals in your body, which support you in sustaining a higher field of awareness. And lastly, an open heart and elevated emotions initiate and support the unification process—the unification of various aspects of being (physical, mental, emotional, spiritual)—and as this occurs you collapse old energies and karmic timelines.

5-D

*My mind is luminous, and I perceive
the luminosity of Spirit everywhere.* [34]

~

The fifth dimension holds exquisitely pure light energy and vibration. Your emotional and mental bodies are clear, as old beliefs have been released and replaced with this pure, higher, lighter energy—an energy that has established new ways of being and living. Cellular changes are commonplace, as are DNA shifts. It is like the rheostat on your genes has been turned up! Innate capacities that were once hidden are now alive. Instantaneous healing is the norm. This is a very light vibrational state. Being in visible or invisible form is your choice, for here you know life is eternal. In this realm, you consciously connect with all dimensions of the universal and cosmic realms. This is the "New Earth" of *Revelation*.

The fifth dimension is a light-filled, dynamic, vibrant state of aliveness and well-being. It is a realm where you know that no one and nothing is against you or can ever be against you. How could they be? For in this dimension, there is no Velcro of receptivity for resistance. Only ease and grace are alive here. There is nothing to fear, for what can harm you? There is no death! There is only a translation of energy from one dimension to another, and you are connected to all dimensions.

A fifth-dimensional state of being, is a fully surrendered state, and thus, holds <u>no</u> vibrational alignment, nor implicit agreement, for the existence of fear energy, judgements, or attachments. Not to jobs, people, control, outcomes, needing to be seen as this or that, labels as identity, old values, situations, money, being safe, persecution, power needs, anxiety, etc. These types of fear-energies cannot, and do not exist, in this dimension, for there is no vibrational alignment with, nor agreement for their existence.

The fifth-dimensional consciousness knows that abundance is the natural state of the universe and is therefore not an issue. How could it be? You have the power of instant manifestation. Every cooperative component you require to manifest all possibility is available to you instantaneously. Energy is malleable and plastic. You know how to work with it in integrity for the highest good of all involved.

The fifth-dimension energy sees easily through all levels of dimensions, for the third eye is clear and fully open. With ease, you read

---

[34] IBID

the energy of a room, group, or person when tuned into them. Thus, you read the energetic records of the heart and mind through a highly tuned sense of intuition. This open intuitive facility has the capacity to see/sense what is coming next.

The fifth-dimension energy holds interconnection, interdependence, and cooperation as key elements, and the highest good of all is paramount and understood in a new way. This fifth-dimensional awareness recognizes resistance and resistant energy immediately but does not engage with it. By being in a state of awareness of conscious neutrality, you have the capacity to consciously neutralize the resistant energy and the energy shifts by simply being in your presence. All things not created from an awareness and choice of truth (things such as manipulation, lies, guilting, control, etc.) are immediately recognized and, not being in alignment with the presence of fifth-dimension energy, are energetically reformed or removed.

In fifth-dimension awareness, all bodies—spiritual, emotional, physical, and—are integrated and functioning as one through clarity of thought and feeling. New values come alive, such as generosity, generosity of spirit, compassion, caring, kindness, integrity, clarity, transparency, authenticity, respect, honor, and inner harmony and resonance. All of which is transmitted through an open heart—some call it love; in higher dimensions, it is known as BE-ing in harmonic resonance with the Universal Impulse.

In fifth-dimensional fields of energy, the following are commonplace: your commitment is to be in service to humanity and, thus, your work is service; being in service to your higher purpose fulfills all your requirements and needs; your soul qualities and talents are active and in service to a greater good; instant manifestation of desires abounds; conscious partnering occurs, both on the inner and outer planes (the inner occurs as divine union and the outer occurs as the capacity to consciously partner with others to co-create higher possibilities for the good of all); an awareness that sees only wholeness and not a need for healing; and an awareness that knows the difference between breaking down and being broken.

In the fifth dimension, growth often emerges through dream states. You start to see inter-dimensionally. You are energetically activated, and your words hold the capacity for activation of others through their sound and tones. You know that you cannot act as savior for anyone and you are not here to save another. You understand that the desire to save another is an ego-oriented impulse that believes it knows best and that trying to save someone is trying to save them from their soul work. Your true work is to support others without judgement and the need to save them or spare them.

In higher dimensional awareness you are able to see through the illusions of separation, and thus, all your actions arise from an elevated understanding of compassion, trust, love, and more. In this consciousness, a compassionate engagement as an inspired action, with someone in need might look like this:

I see your journey; you are working to open your heart and connect with your soul at a deeper level. I will not, cannot, interfere with your experience. Here is something to wear or eat. Energy doesn't lie! You cannot change the world, you can only change you; and as you change you, you change the world—by virtue of your consciousness changing. It can be NO other way!

In fifth-dimensional awareness, you know that you are a soul walking in a body, not a body with a soul. Thus, you see the purpose in all things and make choices moment to moment to observe and refocus on fifth-dimension thoughts and ideals—for you know that you cannot take third-dimensional programmed relationships or ideas into this dimension. In the fifth dimension, your energy has the capacity to neutralize negativity and for it to happen instantly. You are conscious neutrality in action!

Example B is a map of dimensions, a mini overview of what third-, fourth-, and fifth-dimensional paradigms (patterns or models) and their constructs (ideas, beliefs, perceptions, or theories held within the paradigm) mean in relationship to consciousness and life in general as your energy progresses through the dimensions. This chart will support you in viewing your life from a different perspective.

Review each column, starting with the third dimension. Look for the energetic differences in each dimensional paradigm as it expresses itself in the world and in your life. Ask your inner guide to support you in seeing those areas that are your next steps of soul growth. Take notes on your discoveries.

Example B: Map of Dimensional Constructs & Behaviors
3-D, 4-D, 5-D Constructs & Behaviors

Note: a construct is any idea, belief, perception, or theory held within a dimension's frequency range.

Third-Dimensional Behaviors & Constructs	Fourth-Dimensional Behaviors & Constructs	Fifth-Dimensional Behaviors & Constructs
• 3-D reality is fear-based, success-seeking, "me," more, often unaware, unaware of role of TRUE SELF - I AM • Old Earth energy that was prophesied to pass away • Works on the level of the Law of Cause & Effect	• 4-D reality moves from seeking to conscious living. • Aware→ awakening → awake→ alive • 4th D. is bridge between 3rd & 5th D., thus is pivotal and not necessarily stable • 4th D. is path to 5th D. • You cannot do a spiritual bypass on any of the work required • Use of the Law of Attraction & the Law of Causing Effect	• 5-D reality is I AM ONE with living New Earth energy • BE! • Alive → ablaze with luminous energy • Master of energy • Mystery • Unknown/all known • Full surrender to Universal Impulse, the impulse of the ab-SOUL-ute • The Law of Instantaneous Reflection
3-D.	**4-D.**	**5-D.**
• Manifestation is slow, takes time, energy sluggish, dense, is always changing. 3rd D. is impermanent, even though people believe it is permanent; thus they fear and resist change. • What can I get for what I do, often "gimme!" • Seeks outside of self for kindness, compassion, love, worth, success • Dependent on others for good, love, recognition, status, money • Often lives from lack consciousness, even if they have money and possessions.	• Think and many things appear shortly • Testing consistent giving and tithings to see what happens; still often to receive • Seeking relationship within • Tolerance growing for other views and mistakes • Expressing love, compassion, kindness • Open, yet lives with power struggles • Decreasing need for drama/pain body to be engaged • Conflict management desired and worked at, looks for peace • Basic understanding of spiritual journey and what is to be met • Desires to live from "what IS, IS!"	• Instant manifestation • Alchemical transformation occurs in their presence • All needs are provided for without worry • Knows mystery of Universal Impulse, energy, vibration and frequency, resonant fields, body as field of energy, and their relationship to NOW moment • Knows how to fully work/shift energies • Knows energy of ARC of expression in giving; can't out-give Source • Knows the interconnectedness of all life • Lives consciously and connected to the energy of the unified field of consciousness (UFC)

3-D	4-D	5-D
• Low tolerance for other views, mistakes, etc.	• Invites sacred service and greater giving as a constant into awareness	• Knows that in higher awareness, Christ consciousness, there is no fear, no history, no story
• Engages in power struggles	• Spiritually and emotionally evolving	• Sees only one presence, abundance, and wholeness
• Trigger words close off/down—Jesus, Christ, required, responsible, etc.	• Engages daily spiritual practices	• Capacity to shift DNA and recognizes when light is luminous or dimming
• Believes world is against them	• Working "stuff," not through it yet	• Understands 5th D. imperatives, infinite possibilities, dimensions, parallel realities, multidimensional aspects
• Spiritual journey and what is to be met not understood	• Optimistic, positive, working to see from 4th D. metaphysical perspective not 3rd D. perception	• One with UFC, tunes in inter-dimensionally
• Judges by appearances and denies their part in creating relationships	• Works to make conscious choices and take conscious actions	• Works with spiritual masters and guides
• Out to get, not necessarily to give	• Authentic communication becoming important but still often steeped in my need, my wants	• Sees auric fields of color
• Multidimensionality not believed, ego control imperative	• Engaging eternal life concept but still often a fear of death	• Matured spiritual ego, I AM, adverse ego overcome
• Suppresses feelings	• Adverse ego hold is shifting	• Cooperative components and power required from the UFC are instantaneously available
• Ego anchored in me	• Engaging meditation, stillness, silence	• Can often do the impossible, no fear of anything
• Do it for me, I can't	• Bleed-thru of past lives	• Has energy-shifting ability
• Must work for money	• Focused, intentional, accountable, open	• Reads energy in space and can shift the energy
• Blame, shame, I need to know why?	• Exploring, will take risk	• Instantaneous results occur along with suspension of time
• I must feel safe	• Becoming more comfortable not knowing	
• Newtonian cause/effect predicts and controls; linear	• Defining self by spiritual parameters	
• Developing or no spiritual practices	• May experience health issues	
• Attached to my way, my outcomes as they want them; "my will be done"	• Duality-One	
• Fearful of anything unknown		
• Fear of death		

3-D	4-D	5-D
• Fear not recognized as leading life and thus, not worked on • Optimistic, pessimistic, drama, my way, my will • Pain body engaged and lived from • Perceptions rule, sees from 3D. perceptions • Makes unconscious choices and takes action from there • Non-authentic communication; has a need to be right and be sure they are seen as right; thus, conversations often turn negative • Adverse ego is well established • Often fearful of stillness/silence • Often lacks clarity and intentionality • Does not want to be held accountable for anything • Not knowing is very uncomfortable • Fulfilled/defined by world view • Actions motivated by hidden motives • Stuck energy fields • May experience a divine interlude but often ignores or runs from it • If a past life bleed-through occurs, not understood, feared	• Knowing one presence, one power (1P-1P) is all there is • Actions motivated by something greater, often not necessarily only for personal gain • Understands intellectually, but full knowing not yet grounded • Conveying a level of spiritual wisdom, intuitive insight, etc. that begins cutting through stuck and shadow energy • Experiences moments of divine interludes • Works forgiveness on multiple levels • Desires to live in non-attachment, non-judgment • Realizing everything is impermanent • Need for control and manipulation decreasing • Joy in discovering there is nothing to overcome, just opportunities to grow • Clarity, integrity, and accountability of words and actions are invited into play (and still feared) • Willingly participating in shadow processes; shadow aspects arise, are reviewed and questioned • Words and actions beginning to be	• Clarity regarding abundance; all needs provided without taking thought • Clarity, intentionality, accountability, I AM; integrity of words, actions • Knows "I AM holy" as is "All IS holy!" (Titahquah) • No shadow; knows shadow is illusion and is here to teach who we are and who we are not • Has made peace with "Jesus" and all past religious influences • Understands the hidden blueprint in scriptures as a guide for soul evolution • Understands Book of Life and accesses it • Universal law of dominion lived with I AM as fully as dominant vibration of being • Dances at the edge of mystery, unknown, not knowing; unlearning invited and lived into • Comfortable in midst of discomfort • Immortality known, no fear of death • Activations and initiations are spontaneous • Understands universe is constantly speaking through synchro-

Column 1	Column 2	Column 3
• **3-D**	coherent match, but words often still ring hollow when spoken	divinity and synchro-divine events to create "success" on our spiritual journey and that S.D. is a sign of resonance with UFC
• Controls and constructs of mind firmly in place, as are perceptions	• Often engages in metaphysical malpractice and meta-fizzling; i.e. speaks words that are not yet embodied as a living knowing	• Recognizes and bypasses "controls" of mind
• Drama feeds pain body, which is alive and in place	• Landscape of current reality shifting; often is willing to look, see, tell the truth and shift to something more interesting	• Higher good of all is paramount
• Duality-oriented	• Experiencing déjà vu, past lives, dream-scaping	• Expresses from transparency and authenticity, invoking law of dominion
• Shadow not understood and feared; not looked at; it's always someone else's fault	• Shedding old habits as new values arise	• Knows everything is impermanent
• Metaphysics feared or toyed with, not engaged in a way that would change the landscape of literal reality	• Understands roles of archetypes/polarity	• BE and lives are with little or no reaction
• The calculating self firmly in place	• Beginning to understand spiritual journey and obstacles	• Claims, holds, and vibrates in resonance with claim
• Not open to feedback and gets upset when feedback given	• Symbols, signs, etc., are more important	• Knows Kingdom of Heaven is at hand
• Plays games to win	• Dances between dimensions; energy not yet stabilized	• Holds laser-like intentionality
• Not willing to unlearn, etc.	• Divine masculine (light, knowledge) activates divine feminine (the purifier) this is required to support the break down old beliefs and structures	• Experiences interconnectedness and isolation
• Unproductive habits not recognized	• Conscious desire and working to cleanse the past & new sensitivity to energy emerges w/ new skill sets, gifts, & values to use with new life purpose	• About sacred service, not money or fame; spiritual ego alive
• Shadow aspects are intact – the neglect, abuse, co-dependency, & loss you experienced influence your reactions		• Silence imperative for restoration
• Distortion and separation play out in physical body as disease, addiction, me-orientation, etc.		• Not all bliss; understands this and has no attachment to anything showing up any certain way
		• Lives non-judgment, non-attachment, non-resistance
		• No judgment, therefore, no forgiveness of anyone required

3-D	4-D	5-D
• 3-D • Emotions are heightened during "church" & prayer, feels like an experience of God, and yet is often more emotional-oriented than spiritual, yet it feels good • NOTE: We are spiritual, vibrational beings inviting remembrance of why we are here and what we are to do here!	• 4-D • Distortions of reality being seen; rewiring • Let go need to know • Non-linear • Deep soul work being done, diving deep to expand consciousness • Trusting, honoring self • Ego breaking down • Detoxing of cells • Light body being activated, DNA being restructured • Unconscious programming is decreasing rapidly • Fear is a sign to look, not run; karma collapsing • Heart open to higher levels of awareness • Synchro-divinity & spontaneous right action emerge • Metaphysical divine masculine/feminine aspects understood • Sexual energy begins to move from lust and need to a desire for a "need to seed." • New realities are seeded through your creation energy • Phenomena may or may not occur-lights, visions, masters, etc.- as part of "breaking-down" process • Gratitude, acceptance, & trust open the way to higher awareness	• 5-D • Clarity of multi-dimensional and inter-dimensional seeing provides clear energy for viewing life: past, present, future • New Earth realities • In service to humanity • Time ceases • Cellular body clear • New values in place • Secure in self • Unity consciousness is integrated • Whole not healing • Creation energy impregnating body with new realities • Intuitive messaging is the norm • Alchemical transformation of energy occurs in your presence • Live in sacred union, unity consciousness • No phenomena • You are a soul walking a body

4

Conscious Neutrality

[Enlightened is the] ... one who is in unity with the Spirit ...
And who in the midst of the storm,
despite the raging of humans in the chaos of material thinking,
can enter into the silence and
declare the absoluteness of his own being!"
Ernest Holmes

It was dream-time. I was sitting in a classroom, made of pure crystal, and the facilitator was an elder, androgynous, wisdom-teacher. S/he was wearing a flowing luminous robe that shone so brightly, that it seemed to electrically charge the atmosphere in the room—or maybe it was the crystals. It was an advanced teaching session and the ideas being generated, were new to me. At one point, the being looked at me, and telepathically shared two words; conscious neutrality. At the moment of receiving those two words, I awoke. At first, I had no clue as to what the words meant. Yet, every time, I thought about those two words— conscious neutrality—something energetically stirred in my soul.

I realized that I was dancing at the edge of mystery with a new energetic idea, vibration, and wisdom. I felt it, I sensed it, I knew it! I also knew these words, with their hidden wisdom messages, came to spiritually mature and evolve me. My work was to unravel their deeper meaning and learn the lessons hidden within them.

Overtime, as I consciously surrendered into their wisdom, energy, and meaning; the the higher vibrational energy encoded within those two words transformed me; spiritually, emotionally, physically, mentally, systemically, and dimensionally. They also expanded in meaning, and translated into an organized, systematic, conscious, and organic spectrum of processes of learning and unlearning.

The processes then evolved into an instructional design, a curriculum, that supported the releasing, breaking-up, and dissolving of old patterns of perceptions and beliefs; and at the same time an absorption and integration of a new level energy. The new energy now consistently calls forth an interior field of harmonic resonance, inner peace, and state of being that prevails in the midst of discomfort and chaos, even at a level 10.

I began to realize that all of my experiences of working with groups in the appearance of conflict were fodder for understanding conscious neutrality. For in facilitating the various aspects of conflict management, I was now observing the energetics of how chaos arises; how quickly the energy of little problems and issues can escalate; how the energy of anger fractures relationships and organizations; how the energy of factions creates a negative-oriented egoic fervor; how this energy created by and living in the factions can then move at the speed of light into higher degrees of dysfunctional third-dimension-oriented behaviors and conduct; how the energy of unskilled behaviors and actions can affect relationships; and that once the energy of conflict escalates, it is not easy to recreate harmony, resonance, and alignment within a system or organization.

I began to notice that, even, if it appeared that resolution occurred, from a third-dimensional perspective, someone "won" or got their way, it wasn't a real win. Why? Because invariably the undercurrents of the mounting negative energy impacted people on multiple levels of being. They were changed and an innocence was lost; never to be reclaimed again!

Over time, having shifted into being more sensitive to seeing life and its experiences through the lens of energy, vibration, and frequency, I began to sense the vibrational context of an experience, gathering, or event before it was evident to the group. I also became more and more aware that conflict was not about anyone being right or wrong; it was about the energy of the choices being made and how much energetic residue those choices would render.

Observing, that all choices had natural energetic outcomes, and consequences, eventually, my skills for reading energy became honed. I could energetically sense what the results or consequences of a choice would be, at the time the choices were being made. However, I did not and could not interfere. I could only state what was being perceived; the choice for action is still up to the participants!

It became clear to me that people often, make choices that are steeped in inadequate information, and/or in perceptions and meanings they have created about something or someone—and which are not necessarily true, to anyone but the group or person thinking them. Perceptions are, more often than not, anchored in past experiences—not—what is occurring now. Occasionally, I would observe that a person's attachment to their perceptions and meaning-making, along with wanting to be seen as right, created a sense of feeling justified to stay engaged in the conflict and battle—without regard for outcome. Regardless, of options offered.

Eventually, I understood that the appearance of conflict had nothing to with an issue, a problem, or a person—it was always about the level and dimension of energy that a person or group was choosing to participate from in the moment. Change the energy, shift the vibration, and the appearance of chaos will transform.

Additionally, I was discovering and unfolding the subtle nuances of what it meant to be: a master facilitator of energy in each moment; an alchemical transformer of energy, mine and others; the presence of all is well in the midst of chaos; be in service to change and transformation; aware of what conscious neutrality was, and what it was not!

I realized that conflict was not so much about what appears to be the issue, as it was about energy being out of harmonic resonance and alignment. With that in mind, it was easy to begin to envision the system (the people and the appearance), as an energetic puzzle where pieces have popped out of place. My work was to support them in the creation of a new picture or vision.

The understanding of a simple quote, "*If you want to know how the universe works, think in terms of energy, vibration, and frequency,*" over time, invited me into a whole new way of asking questions. The energetic shift into asking more empowering questions began to reveal its effect in both my personal energy field, and the field surrounding me.

In other words, my growth was that I moved from a third-dimensional way of seeing conflict and its management, into a fourth-dimensional (and higher) interpretation of the energetics involved. This shift in awareness invited both a new way of seeing the experience and a new understanding of the types of questions that were important to ask, in any given moment or situation.

I began to ask empowering questions like, "What might be more energetically interesting here? What new energy, or opportunity, might be desiring to emerge?" I asked the participants to engage honest self-observation, reflecting on, "How are you contributing to the chaos? What is one thing you can do evoke change, in this situation?" o

Viewing situations (appearing as conflict) from this energetic-vibrational perspective increased my capacity to dance at the edge of mystery and energetically sense what was occurring within in the field of the group energy. As this shift occurred in me, the way I engaged perceived conflict situations also shifted and people felt it. More and more, tense conversations quickly began to transform in the moment.

> *Because the sage always confronts difficulties,*
> *he never experiences them.*
> Lao Tse

~

Prior to discussing, in detail, what conscious neutrality is, let's review what conscious neutrality is not. Conscious neutrality is *not*:

- A state of pretending to be at peace and yet simmering interiorly at near boiling point.
- A contrived state of awareness in which you try to talk yourself down or out of something: "*Oh, if I stay calm, stay calm, stay calm, stay calm, things will be okay, I think!*"
- A pseudo-community in which we all pretend to be in a wonderful family relationship and yet we only tolerate each other.
- A pseudo-relationship based on the idea that if we can just keep people from saying anything too authentic about how they really feel, everything will be okay.
- Halting what is occurring in the moment by rushing to call for prayer so that any uncomfortable feelings being expressed may come to an end. After all, who can stand the discomfort?
- A state devoid of emotion, where nothing gets expressed or spoken.
- About fixing problems or issues at the level in which they occurred.

If the above is what conscious neutrality is not, then what *is* conscious neutrality? Conscious neutrality is a natural fourth-dimensional state of being that, by its very nature, invites a more highly-developed, natural, spiritual and emotional response to situations, people, and perceived discomfort.

Conscious neutrality, as a state of mind and consciousness is initiated as you are willing to say, yes, to taking on more of a fourth-dimensional awareness, and higher. However, to reach this higher state of awareness requires that you consciously make choices that will dismantle old and outdated third-dimensional perceptions, beliefs, feelings, etc. Thus, conscious neutrality supports you in increasing your capacity to not judge by appearance, to not judge through the lens of past perceptions, beliefs, and embedded emotional reactions. It does not mean that you do not care, or that you ignore what is going on around you, it means you do not judge. For you learn that judging through the lens of the past only increases and amplifies the amount of dense energy in your field and that old or past energy, cannot make a change in your field, or the field of humanity, you can only make lasting change through the energy held in the present moment.

You, by virtue of being human, and a part of the collective consciousness, are an organic structure. As such, you enjoy your embedded patterns of behavior and response. You, unconsciously, enjoy reliving those old patterns, and shadow aspects, for they are emotionally comforting. For you can say, with emotional satisfaction—see I told you so, this always happens to me.

As you engage the journey into conscious neutrality, 4th dimensional awareness, you begin to understand that reliving these old patterns, and then judging your life from the emotional residue of those patterns and shadow aspects, assures that these patterns will show-up in other areas of your life. And as do so, they create more dense energy in your energy field. Thus, is why you make the choice to engage in a conscious dismantling process. You do this dismantling until you have rewired a new norm; for to achieve the state of mind and being that holds conscious neutrality as a consistent and established norm, requires a neural repatterning. Neural repatterning affects the entire being—mind, body, spirit, soul.

In fourth-dimensional metaphysical language, this repatterning and rewiring of the neural pathways is known as, establishing a demonstration in mind that is both sustainable and repeatable. Know that even after the demonstration is made, old perceptions, thoughts, beliefs, and feelings can and will try to sneak up on you and try to catch you unaware.

The development of conscious neutrality initiates the development of a systematic, organized, and progressive demonstration of conscious soul growth, spiritual principle, spiritual and emotional maturity, movement into being a master facilitator of energy. Mastery is reflected by the consistent and practical application of spiritual principles; consciously making choices to live from elevated emotional responses; the development of higher-level values and qualities;

inspired action; and conscious application of all these in everyday interactions and decision-making processes.

Inspired actions, are actions inspired by a higher dimensional awareness, rather than arising from a 3rd dimensional-oriented, reactionary state. Inspired actions are spiritually informed, hold spontaneous right action, and a state of non-attachment to outcome. And, their intention for outcome, always supports the highest and best for all involved.

Inspired actions are empowered by elevated emotions. Elevated emotions arise from an entrained, coherent heart-brain connection. This coherency then elicits higher-level feelings and elevated emotions such as, trust, authenticity, integrity, appreciation, joy, transparency, love, peace, etc. Thus, conscious neutrality is a state of being neutral, but it is not a state devoid of emotion.

Being in a state of conscious neutrality does not mean that you do not care, or that you never get mad, sad, angry, or express any emotion. It means that you do not carry the emotional baggage of past experiences, for you have evolved to a state of being where you have command of your emotions. It is the ability and capacity to hold any emotion in a neutral, consciously-mindful, elevated state while offering and inviting Higher Self-Spirit-Source Energy to constructively express through the emotion.

You know how to feel your emotions and express them constructively, rather than your emotions having command over you. You can observe a feeling rising, such as anger or sadness, acknowledge the presence of the feeling, and make a conscious choice to express it in a way that is constructive and keeps the energetic space open. This energetically supports you in acknowledging the emotion—not denying it—and as seeing the experience as a teacher and as having purpose for a greater good.

Conscious neutrality, as a developed state of being, holds an established consciousness of abundance, and thus, creates the energetic space for the greater good of all, not just for self.

Conscious neutrality invites and leads to a systemic integration of higher dimensional frequencies as a current, living reality. Integration, absorption, and assimilation of what you are learning and living creates an expansion of consciousness and conscious awareness. It is this expanded awareness that supports you in reading energy and in viewing situations, people, emotions, and energy from the lens, or vantage point, of 4th, 5th, and higher dimensional perspectives; thus, allowing you to see experiences and reality, differently.

Conscious neutrality, as a soul evolutionary process and an established state of being, affects, transforms, and shifts the energy, vibration, and frequency within your physical, mental, emotional, spiritual, and auric fields, resulting in the capacity to stay present in the present moment. Remember, it is only as you are present, in the present moment, that you can read the energy in a field—yours or others!

As conscious neutrality in action, you effect the energy in a room or situation, for by virtue of a higher consciousness connection, you have the capacity to expand an energy field (resistance contracts a field). Expansive energy alchemically alters situations through raising the vibration in a field. Thus, holding expansive awareness has the capacity to energetically dominate a space and initiate new results. In the expansive vibrational field of conscious neutrality, which holds spontaneous right action, walls collapse and fear and anger dissipate, and it is done without fight or protest.

Conscious neutrality is developed through practicing and engaging the tools of honest self-observation, nevertheless I am willing, spiritual mind discipline, surrender, meditation, authenticity, conscious communication, transparency, etc. All of these are spiritual tools that support you in seeing when you are in alignment or out alignment. And if out of alignment, they support you in making choices that will consciously realign you. Consistently, doing this, overtime, will result in a remapping and rewiring of your neural pathways and establishing conscious neutrality as a state of being that is sustainable.

As energetic and alchemical transformations occur in your consciousness and being, your energy field will become luminous. You will radiate with a luminosity and an essence that arises from the energy of the universal presence shining throughout you and your energy field. The authority, power, and dominion of universal presence—I Am—is alive, aligned, and attracting through you, and is now the inspirer of all the actions you take.

Your actions arise from a new level of consciousness, one that demonstrates who you are! *By their fruits they shall be known!*

Quantum leaps...are evolutionary drivers...
they are vital stimulants which trigger
astounding design innovation.

Barbara Marx Hubbard (paraphrase)

~

Things are not always as they seem!
There could be more to this than meets the eyes!

~

Although, the words conscious neutrality came in a dream, it was a nudge from the Universal Impulse, that really opened the way for deeper understanding and awareness of what they meant. Through an act of synchro-divinity, I discovered four contemporary-oriented key elements hidden in the secret wisdom and mysticism of the *Sermon on the Mount*. Understanding the metaphysics and mysticism of these four key elements—related to the Law of Attraction and the Law of Causing Effect—supported me developing a deeper understanding of the ideas underpinning conscious neutrality.

The four key elements are:

BE = Align + Allow + Attract

~

BE = Align + Allow + Attract, when unraveled through a contemporary metaphysical interpretation, invite you into an a deeper understanding of the following: how to be conscious neutrality in action; how to be in alignment to higher dimensional vibrations and frequencies within your entire body cellular system; and how-to create the shifts in consciousness that are required to consciously attract new dimensions of awareness and to evolve a consciousness of soul mastery and the mastery of abundance in <u>all</u> areas of your life.

However, know that movement into any type of higher alignment, often shakes and shifts the core of your foundation. So, get ready for shifts in your attitudes, values, levels of expression in relation to spiritual and emotional maturity, ways of being, and more. And, remember, there can be no spiritual bypassing—you must do the work— for it is the dominant vibration of your being that determines the level of frequency to which you align, and from which you attract possibility from the unified field of consciousness.

As you read about the four key elements and then engage the detailed metaphysical teachings that the whole *Sermon on the Mount* holds (refer to the chapter on *Soul Evolution*), stay alert and watch for how these key elements reveal themselves, not only as stepping-stones to conscious neutrality, but also in the development of the spiritual principle of abundance and the law of attraction:

BE: <u>MT. 5: 1-20</u>.

Beatitudes: BE, who you BE as conscious neutrality in action, is the *end result* of practically applying the metaphysical and mystical message found in the *Sermon*. It occurs as you:

* BE—who you BE—as you engage a new state of consciousness and conscious awareness; referred to as conscious neutrality.
* BE the establishment and demonstration in consciousness of fourth- and fifth-dimensional principles and information, which creates new awareness' through an alchemical transformation of BE-ing.
* BE and know that you are a point of power in the universe. Why wouldn't you create and attract what you want? You are the point in the universe that Universal Presence works through.
* BE in harmonic resonance, in consciousness and at the cellular level of your body system, with higher dimensional levels of energy, vibration, and frequency.
* BE a walking field of higher vibrational energy that is both felt, observed, and perceived by others that encounter your field.
* BE in coherence, consciously align with elevated thoughts and elevated emotions, for they create a coherent and dominant field of energy. That field of energy is known as a vibrational broadcast signal (VBS), and that signal goes forth to attract back to you, in accord with your VBS, the highest possibility in the unified field of potential. You BE abundance in action!
* BE, through practical application of spiritual principles, conscious alignment, and conscious sustainability of the teachings found in the key elements of Align, Allow, & Attract (MT. 5:21–>MT. 7:29).

ALIGN: <u>MT. 5: 21 —> MT. 6: 24</u>

Align/Alignment initiates the inspiration and metaphysics of the teachings.

* Aligning consciousness with higher dimensional frequencies of awareness happens initially through prayer, meditation, self-observation, forgiveness, letting go of all anxious thoughts, and other conscious practices. Once aligned with these higher frequencies, you must then learn how to allow them to do their work in and through your energy field. Allow them to create an alchemical transformation in and through your being, on all levels—physical, mental, emotional, and spiritual. This clearing will be done through the choices you make, as you live your life experiences.

ALLOW: <u>MT. 6:25 —> 7: 6</u>

You allow by being a field of nonresistance (energy).

* Non-resistance is developed by being willing to make the choice to unlearn everything you thought you knew, by being open to doing the deeper inner clearing and cleansing work presented to you to through your life experience, and by being in constant and conscious surrender to Source Energy.

* Remember, Source Energy is an ever-expanding yes vibration and, as such, doesn't work in terms of big or little ideas. It works energetically and vibrationally in terms of how much resistance or nonresistance is present in the field in which It is working.

ATTRACT: <u>MT. 7: 7-29</u>

Attraction is what happens to you, naturally, as you engage fully with the energy, vibration, and dynamics of aligning and allowing.

* Through the new fields of energy, you are creating, through aligning and allowing, you create a new vibrational signal and a new level of vibrational momentum to attract from. This new vibrational signal and its momentum begins to attract a higher level of possibility from the unified field of consciousness. However, attraction is not enough: you must be willing to take inspired action on that which is attracted to you. Action rendered through the inspiration of Source Energy, brings new possibilities to life.

As you study the metaphysical and mystical key elements underlying the *Sermon* (and the whole soul evolution template found in the *Gospels*), allow their energetic meanings to expand your consciousness, and then begin to practically apply them in your life; your life will change dramatically, you will alchemically transform, as you take the time to absorb the energy and meaning in these key elements. Don't rush it!

The mysticism shared in the *Sermon* is a vibrational transmission and a mystical treatise for those who have *eyes to see*

BE = ALIGN + ALLOW + ATTRACT

Attraction supports action that is in
vibrational alignment with who you BE!

A&A: The 33 Tenets of Conscious Neutrality

One who is in unity with the Spirit, is majestic in his power...
[is conscious neutrality in action],
who in the midst of the storm, despite the raging of men in the chaos of material thinking, can enter the silence and declare the absoluteness of his own being!
Ernest Holmes

These 33 tenets are the foundation and underpinning of *Conscious Neutrality;* a natural 4th dimensional state of being. The tenets are not, by any means, an exhaustive list. They are just meant to be touchstones. However, when you consciously choose to establish a demonstration in mind of any of the tenets or principles, you will find yourself in a more expansive state of awareness. For you will discover that through the choice to engage in conscious transformation, you have deconstructed old paradigms and ways of thinking and feeling, and rewired new paths of neural networks.

Several of the words in the tenets (indicated by *italics*) are gleaned from a metaphysical-mystical interpretation of the *Sermon on the Mount*. It was the unraveling of the *Sermon's* hidden wisdom that revealed the qualities, behaviors, and values that one must align with in order to live from the 4th dimensional state of *Conscious Neutrality.*

The tenets include, but are not necessarily limited to the following actions, qualities, values, practices, and behaviors:

1. *Peacemaker* — one who has tapped into and aligned with a field of harmonic resonance. A field that holds the vibrational energy of the spiritual principles of peace, joy, love, etc.
   a. Having tapped into this field of harmonic resonance you have the capacity to still your emotions on command, and to consciously elevate your thoughts and emotions in the moment; so that higher level decision-making processes are engaged, and spontaneous right action is taken.
   b. You ensure that all your decisions support the greater good of all, not just the good desired by the 3rd dimensional adverse-egoic self—yours or others;
2. Compassionate-allowing
   a. Compassionate-allowing arises from your being willing to BE; *meek*/humble, *poor in spirit*/egoic-will surrendered,

merciful/forgiving, *pure in heart*/authentic, transparent, and non-attached to outcome.

3. Aligns with Source Energy, allows It to lead and to attract the highest possibility! *Seek first the kingdom & all will be added.*
   a. You *hunger* and *mourn* (yearn) to have Spirit as the only presence and power in your life. Thus, you are willing to consciously surrender, in every moment, to Spirit's guidance. *He went to the mountain top and sat down!*
   b. You consciously invite Spirit to be the dominant vibration in all you do and be, and to attract to you the highest possibility from unified field of consciousness, in accord with that vibration! *Go into your closet & shut the door...*
4. Spiritual and emotional maturity!
   a. You allow this level of maturity to permeate all interactions, actions, and all decision-making activities.
5. Willing to speak authentically and with transparency.
   a. You speak and/or take action without a need for boundaries, protection, or the assurance of safety.
   b. You know there is no need for safety in higher awareness, for trust is now embedded in higher vibrational awareness.
   c. You are skillful in the engagement of the art of asking empowering questions and holding conscious conversations.
6. Conscious neutrality in action!
   a. Conscious mindful neutrality. You are not void of emotion, your emotions under command of Higher Self/Source Energy. *You cannot serve two masters!*
   b. You demonstrate the ability & capacity to neutralize negativity & fear in yourself & in others.
   c. You accept views and insights without judgement
   d. You are a generous listener, without the need to give advice or fix another. *Golden rule!*
7. Consistent, clear, and conscious choice-making!
8. Engaging non-attachment, non-judgement, non-resistance!
   a. You understand that you cannot grow spiritually without clarity around the role non-attachment and non-judgment play in spiritual expansion.
   b. For you know that without the capacity for non-attachment you tend to: take things personally; be needy; be ego-centric; be self-identified with your problems; court fear and its

friends; make others wrong; camp in the valley of ain't it awful; collude; possess a great need to be right or to be seen as being right—which alters your decision-making capacities; and much more. *Do not judge by appearance.*

9. Energetically sensitive, can read and sense energy in a space.
    a. You sense and "see" energetically what is, and what is not, present in the field of energy around you.
    b. You sense how to restore harmonic resonance and balance that is for the good of all, not just you and thus, you move forward with choices that reflect higher dimensional awareness. *He could read their hearts!* MT.9:4
10. Collaborative and interdependent with all sentient beings and the environment. *Love is the fulfillment of the Law!*
11. Accountable for living from a solution-oriented awareness.
    a. You know: "I am responsible and accountable for my life, and actions, without any need to project or blame another."
    b. You are you in allegiance to no one or nothing less than Spirit guidance. *Take no oaths!*
12. Accountable for every action you take, or do not take! *Every penny will be accounted for!*
    a. You are willing to own your story, without a denial of a part in the creation of it, and you shift the energy into a narrative.
13. Accountable for the impact of your energy and vibration on, and in, all your interactions, and in the creation of your experiences.
    a. You are aware of the role of energy, vibration, and frequency plays in awakening and in the creation of multi-dimensional awareness. *Blessed are...*
    b. You hold the capacity to consciously facilitate energy to transform self and/or in a space when energetic inharmony, discord, or high levels of discomfort are present.
    c. You are aware of how conscious energy facilitation supports the dismantling of third-dimensional beliefs & perceptions.
14. Spontaneous, flexible, and willing!
    a. You are comfortable dancing at the edge of mystery, sitting in the midst of the unknown, and being in the question.
    b. You are willing to allow the questions to lead, not answers!
15. Capacity to see the vision of a bigger picture.
    a. You recognize paradox, polarity, and the both/and in situations vs. seeing only black/white.

16. Aligned with grace and ease, as a presence.
    a. You are present to the energy in a space without resistance, anger, fear, reaction, attachment, or judgment.
17. Embraces simplicity, without drama and complications.
    a. You surrender, allowing things to happen and spontaneously arise, rather than forcing or manipulating outcome.
    b. You experience freedom, you are free of fear, for you know that there is only One Presence, and One Power.
18. I AM is the dominate vibration of BE-ing.
    a. You allow, through surrender, the I Am energy to be the dominant vibration of your BE-ing.
    b. You engage in spontaneous right action, inspired actions—action that is authentic, transparent, and often calls forth vulnerability.
    c. You are courageous and willing to trust enough to move forward without a guarantee of an outcome—*blessed are those who are <u>willing to be</u> persecuted* (to walk without fear).
19. Allow inspired values to rule—authenticity, authentic communication, intuitive insight, trust, clarity, intentionality, joy, awe, wonder, peace, interdependent, and the capacity to be willing to dance at the edge of mystery.
    a. You are authentic communication in action.
    b. You are clear, courageous, and authentic with the words you speak. You are willing to speak fearlessly and compassionately, with no fear of outcome. Trust reigns!
    c. You ask three important questions, prior to speaking; 1. Does it need to be said? 2. Does it need to said, right now? 3. Does it need to be said, by me? (Bryon Katy).
20. Honest self-observation is your spiritual practice.
    a. You notice and observe yourself and honestly review your energy and feelings, then you take actions to shift any area where the energy does not feel pristine.
    b. Your actions are free of any need to misuse power for selfish gain and are oriented toward sacred service.
21. Positive cultural preferences in action!
    a. You are a state of peace, so, there is no need to react to cultural, societal, political, or family beliefs or issues. You engage social action from a higher perspective.

22. Energetically the field of the presence of "all is well"!
   a. You know that "all is well," and it is not about feeling conditionally good based on how someone else treats or perceives you.
23. Consciously aligned with a higher awareness that knows that life is always for you. *Take no thought*!
   a. You know that life is always revealing who you are, and who you are not, so that you might glean wisdom through your experiences.
24. A contagion for higher vibrations and energy!
   a. You know that in your presence, people catch your energy like a virus. You know that holding a higher-level vibration cuts through negativity, and acts as an alchemical transformer. You know you are that presence! *By their fruits they will be known. Taught as one having authority.*
25. Free of any temptation to misuse power, vomit emotionally on others, or create drama and stories for secondary gains.
   a. You forgive (give love for) as you hold an unconditioned state of harmonic resonance. *Leave your gifts at the altar.*
26. Inspired action is your norm! *You are the salt, the light!*
   a. You are intentional, focused, laser-like in your actions.
   b. You take actions, that are inspired by Source Energy, and that arise from conscious and honest self-awareness, self-observation, and spiritual and emotional maturity; and not from fear, the need to be right, or to be seen as right.
27. Align, Allow, Attract.
   a. You are in alignment with Spirit and higher guidance.
   b. You will not *divorce* from this alliance, under any circumstance.
   c. You are a state of non-resistance and non-attachment to outcome; always attracting an outcome that supports all!
28. Comfortable in the midst of discomfort, even at a 10.
   a. You can shift your awareness in uncomfortable situations.
29. Care but not carry.
   a. You have command over your emotional responses and respond in a constructive manner, as opposed to your emotions having command over you.
   b. You are a state of neutrality, but not a state devoid of emotion. You care, but do not carry. *Do not be anxious!*

    c.   This is not a 3rd dimensional state of neutrality—a disengaged neutrality—as in I refuse to comment or participate for it might make me look bad or reveal true thoughts on issues.

    d.   You engage skilled behaviors vs. unconscious & unskilled.

30. Spiritual principles—abundance, life, and truth—demonstrated!

    a.   You consciously create these are a part of your reality.

31. Creates spontaneous moments of synchro-divinity and synchro-divine events! Right time, right place, right people, it just happens.

    a.   You know that moments of synchro-divinity are the natural consequences of alignment with Source Energy.

32. Creates energetic fields of "right thinking," not "fight thinking."

    a.   You stand for spiritual-social justice through conscious and inspired action. Inspired action holds the capacity for alchemical transformation through your presence of all is well; not fighting or resistance. *Agree with thy adversary, quickly!*

33. BE! The master teacher Jesus, in the Beatitudes (BE-Attitudes), said over and over, *Blessed are those who be...*

    a.   *Bless*, metaphysically means to consciously confer the attracting and multiplying energy of Spirit upon someone or something; to act as an attracting agent for increased good.

    b.   *Blessed are*! Are—is now—completed—done! The energy of Spirit as a multiplying agent and vibrational attractor, is conferred upon—embedded in the consciousness of s/he who has aligned with the energy and vibration of the higher dimensional—metaphysical—mystical—wisdom-teachings contained in the *Sermon.*

    c.   *Blessed are* those who *BE*! Who have done the conscious work and thus, be and live from an established state of 4th dimensional awareness.

    d.   *Blessed are* those who *BE* conscious neutrality in action, in the world.

    e.   *Blessed are* those who *BE consciously aligned with 4th dimensional awareness, conscious neutrality.*

    f.   This list of 33 is not exhaustive, there is always more . . .

5

Soul Evolution

*I believe that what Jesus and Mohammed and Buddha
and all the rest said was right.
It's just that the translations have gone wrong.*
John Lennon

~

*Life is an ever-progressive, ever-evolving,
upward spiral of conscious evolution.*
Charles Fillmore

~

I have discovered that when the Universal Impulse wants me to do something—and I am aligned with and receptive to Its guidance—It is often relentless in Its energetic urging and nudging! And frequently It keeps turning up the energetic heat until I take an action!

It was early spring when I received the energetic nudge regarding reading the *Sermon on the Mount*. At first, it wasn't a particularly dramatic or exciting prod; however, it was audible, and it came as a felt, energetic urge, and a knowing. The message was clear, I was to read the *Sermon on the Mount*. Not being particularly excited about the prospect, my thought was that I would do it in a few days, as I was reading something, I thought was more interesting at the time. But after several minutes of relentless pursuit by Spirit, I knew that I had to do it, now! Within a short span of time, I felt like it was an imperative. Spirit is strange like that, at least in Its dealings with me.

What happened when I picked up the Bible and turned to Matthew was unnerving. It was a moment of expansive consciousness and in the midst of it, I began to see energy moving under the words. As the energy moved and swirled, it began to download and imprint new information and data—within my being. I felt it! I didn't quite understand it, yet, though.

It was an invisible, vibrational imprint of data that held higher dimensional wisdom, knowledge, and meaning, and my work was to recognize its presence and allow Spirit to unpack its meaning through me. As I unpacked the data, I began to see that this ancient teaching of the *Sermon on the Mount* was a mystical treatise on soul growth, and the law of attraction. It was a hidden, encoded, mystical how-to teaching on how to be in energetic mastery, a master facilitator of energy, and conscious neutrality through working consciously with energy, vibration, and frequency.

One of the more contemporary metaphysical ideas that came to for me, was that of seeing that the Sermon held key elements in support the law of attraction, mastering abundance, and the establishment of spiritual principle through neural repatterning. Understanding soul evolution and the various initiations you will meet on the journey is important to your soul growth, if you desire to master, lower dimensional energy, energy facilitation, and live in soul mastery. It is also an important, in that if you can see the path, you can support others in traversing where you have been.

Throughout the centuries, mystics and scholars have recognized that true spiritual sojourners experience similar patterns of soul growth and initiatory sequences. Eventually, these mystics and scholars began to assign names, stages, states, and levels to the patterns and initiations being observed. Those who have shared their descriptive wisdom include Dionysius, St. John of the Cross, Theresa of Avila, the master teachers of the Egyptian Mystery Schools, Charles Fillmore, Ernest Holmes, Carl Jung, Ken Wilbur, Clare Graves, Don Beck, and Joseph Campbell.

Additionally, are you aware that the archetypes, and various other types of symbols appearing in your dreams—like that of going on a journey, interacting with a newborn baby, being caught in a flood, losing a house, or finding a new house—are often pre-initiation indicators? They are pre-cognitive signals that shift, change, or something new is about to occur in your life. So, as you traverse the path of soul evolution, know that you will meet archetypes, initiations, metaphors, milestones, clearings, activations, experiences, signposts, markers, cycles, and metaphysical elements.

Each of these are common elements found in the collective spiritual experience; and yet, each experience you have of them, is unique and designed just for you as a "wake-up" opportunity. Consciously meeting these opportunities and transforming their energy creates the capacity to progress to the next phase of the journey. One such marker on the journey, in modern, quantum language is aligning with the state of conscious neutrality.

The transformational aspect of the soul journey, with all of its many opportunities for growth, requires you to consciously dissolve, release, let go, forgive, surrender, be comfortable in the midst of discomfort, learn to dance at the edge of mystery, realign, remember, and more. At this stage of human development, that takes conscious participation! And what better way to participate than to have a road map or template for easy identification and understanding of the mile markers.

Is the road to soul mastery and spiritual evolution easy? No! Is it a spiritual and energetic imperative to traverse the path of soul evolution through conscious participation in the lessons provided through your life experiences? Yes!

> *The assurance of trust*
> *is not about someone having your back*
> *or assurance that you are safe.*
> *It is about gaining the trust of your*
> *True Self as protector, guide, and wisdom-teacher.*
> ~

The Bible is written in symbol, metaphor, and parable. The key, however, to understanding its symbology, etc. is metaphysical interpretation. Yet, to see the full depth, richness, and mysticism of metaphysical interpretation, it must come through a mind that is open to new possibilities.

For me, the spacious of that possibility blossomed as I was willing to: make peace with my religious past; get over "Jesus as my Savior" resistance; let go of the question, "Did Jesus, even, really ever live?" As all the energetic density of doubt and resistance regarding "Jesus" was released and cleared from my being, a new energy arose. This new state of non-resistance, along with a newly developed sense of trust, energetically opened the way for me to see, with clarity, the mysticism hidden in Jesus' teachings. It was from this state of awareness, that I was then able to clearly see the journey of soul mastery, and the importance of each aspect of the quadrilogy—energy-vibration-frequency, conscious neutrality, mastering abundance, and soul evolution—as a unique, yet, integral component of the journey to mastery.

> *The greatest deception humankind suffers,*
> *is from their own opinions*
> Leonardo Da Vinci
> ~

There is a renaissance afoot that is calling for humankind to step into higher-dimensional awareness and at the same time, to reach deep into its "spiritual" I AM roots. Humankind is also being asked to free itself of its human-created, religious dogma. This renaissance is being called a 5th dimensional awakening, multi-dimensional awakening, Cosmic-Christ consciousness, Spirit, the path of ascension, etc. Mr. Fillmore shares some ideas on this, with these words from the *Unity Correspondence Course, Introduction and Lesson 1* [author's insert]:

"The Science taught in these lessons is founded upon Spirit [higher dimensional awareness]*...It is not necessary that* [one's] *spiritual nature be awakened before beginning this study. The consciousness is quickened by the Word of Truth, and as these lessons are faithfully studied, the living Word of Truth which is in them will enter into the mentality, and the quickening of the understanding will take place.*

The very foundation of all Truth is the right understanding of God...Jesus said, "God is Spirit. "Mind and Spirit are virtually the same. If you know about mind, you know about Spirit, or God...God as Spirit is individualized in man ...an indwelling identity, which seems personal, but has no limitations of personality...God was so fully individualized in Jesus that he could say "I and my Father [Mind] *are one." ...*

God as Principle is the unchangeable Life, Love, Substance, and Intelligence of Being... Different religions have different names for this One...

When we clearly discern the Science of Mind, we will solve all the mysteries of creation... God dwells in us as ... God creates and moves through the power of mind. It is through our minds that we shall find God and do his will. Man has consciousness in the One Divine Mind. When man thinks he has a mind separate from God-Mind he builds a state of consciousness adverse to Truth, which in scripture is called the "Adversary" or "Satan" [Adverse Ego]*.*

All the ideas contained in the One Father Mind are at the mental command of his offspring...The "kingdom of heavens" is the realm of Divine of Mind, or the Order and Harmony of God-Mind, which descends into the mixed mortal minds of men and sets up...states [dimensions] *of consciousness. Truth must first be comprehended and established in man's consciousness..." Thy will be done on earth as it is in heaven." You must make conscious union with it* [Divine Mind]*. The point of contact is willingness, and a seeking...*

Why are we not naturally conscious of its Presence and oneness?...we have used our inheritance, the power to make ideas visible, and created a realm that separates us in consciousness from the Father [One Mind—5th dimensional awareness and higher]*...For Scripture authority on realization of Divine consciousness read the third chapter of John. Jesus' mighty works were done in the consciousness of oneness with the Father. "I and my Father* [Divine Mind/Universal Mind/Universal Presence] *are one."*

The knowledge of God is received as quickening spirit, the whole man is awakened and transformed through it…It must be individually experienced… When we make ourselves one with Divine Mind [Higher-Dimensional Awareness] its Ideas become quickened, established, and incorporated in our consciousness, and we manifest the image and likeness of God which we are…all our ideas must be one with Divine Ideas, and must be expressed in the Divine Order of that Mind."

First-century Christianity teachings, *"Holy Spirit initiated, Christ-based teachings,"* were foundational instructions meant to initiate a renaissance within the people of the time. It was a renaissance that was meant to light the world on fire and create a burst of new energy that would spread like a wildfire. However, only a few caught the flame.

Those first-century persons who caught the flame, did come together to foster the fanning of the flame, through its unique message. It was this deeper realization of Spirit within, a higher dimensional awareness, that was the real purpose for those first century gatherings; not the growing of churches or organizations.

Jesus was an everyday man of the times. He evolved in consciousness, just like you, and he became a teacher and way-shower for the establishment and full demonstration of Christ principle in mind and being. His teachings were revelations from his own inner Spirit, coming through a growing awareness and connection with Source. This spiritual connection once grounded interiorly, through the activity of prayer, was transmitted throughout all of Jesus' interactions to others; those persons who were receptive to higher vibrational teachings.

Mr. Fillmore, in 1923, at the first healing conference presented by Unity®, and held at 9th and Tracy in Kansas City; emphasized over and over, the importance of the spiritual role that the man named Jesus, played in the development and evolution of the mind as a healing instrument, as a healing mind. He believed that Jesus, a human being like you, served as a prototype and role model for the unfolding of the next phase of conscious evolution.

"…Only after the inner life is awakened
is the true sense of the spiritual word understood…
Jesus Christ more fully voiced this than anyone else…
His words are correspondingly vivified with inner life and fire…
The process by which Jesus evolved from sense to soul was,
first, the recognition of the spiritual selfhood and [then]

a constant affirmation of its power...
By the power of His word He brought about the realization.
The mind moves on ideas; ideas are made visible through words.
Hence holding right words in the mind
will set the mind going at a rate
proportioned to the dynamic power
of the idea back of those words.
A word with a lazy idea back of it will not stimulate the mind.
The word must represent swift, strong, spiritual ideas
in order to infuse them with the energy of God into the mind.
This is the kind of word in which Jesus reveled.
...affirm in silence and aloud until the very ethers vibrate with truth,
'I and the Father are One.'
'All authority hath been given unto me in heaven and on earth.'...
Mental inertia of the mind will be overcome and
the way opened for the descent of Spirit." [35]

~

And s/he taught as one having
authority and dominion, for s/he knew;
I am the presence of the living Christ consciousness!

~

As stated, Fillmore believed that Jesus came to earth to evolve in consciousness, just like you are doing. Through the events and experiences of his life, he left a blueprint or template for that evolutionary process, for soul mastery. This process, when understood and seen with *eyes that see*, results in an expansion and ascension in consciousness from a third-dimensional reality expression of consciousness to an expansive connection with multi-dimensions of frequency, knowledge, light, and awareness of fourth and fifth-dimensional energies, and higher.

Through his evolution, as shared in the Gospels, Jesus embedded in collective consciousness, a template for how to demonstrate the expansion of consciousness, also known as Christ consciousness, fifth- dimensional awareness.

It is known that the master teacher Jesus, shared mystical wisdoms and teachings through parable, mystery, and story. He left those hidden, mystery teachings for those who were willing to discover and engage in the process of unlocking the hidden messages and codes. For many centuries master teachers and students of theology,

---

[35] Charles Fillmore, July 1923 Address, *Unity 1st Conference and Healing Revival*, 9th & Tracy, KC, MO.

mysticism, and metaphysics have been working to unravel these codes and mysteries.

> *"My religion consists of a humble admiration*
> *of the illimitable superior spirit who reveals himself*
> *in the slight details we are able to perceive with our...mind."*
> Albert Einstein

~

In a 1930s talk, Fillmore described the spiritual journey as having seven distinct phases or initiations. These initiations are in alignment with and relate to the experiences you will encounter on your soul's journey. The initiations are a foundational underpinning for the Christ template Mr. Fillmore discovered hidden in the life and teachings of the master teacher Jesus.

The importance of the initiations is that they serve as a guide for soul growth and evolutionary awareness. If you know and understand that there are certain initiations that are part of the spiritual journey, you may be able to recognize when you are in the midst of one. This recognition could be used as catalyst to support you in moving through the initiation and learning your lessons more easily. The seven initiations, as purported by Fillmore, include:

1. *Birth*
2. *Two Baptisms (Water and Spirit)*
3. *Wilderness-Temptations*
4. *Demonstration through Ministry Leads to Transfiguration*
5. *Gethsemane*
6. *Crucifixion*
7. *Resurrection-Ascension*

Each of the seven initiations has distinct markers, signposts, and milestones of achievement related to the soul's journey and the continuation of the process of soul unfoldment. Fillmore also shared that within each of the seven major initiations, are multiple minor initiations. So, it is not seven initiations and you are done, but each is akin to 7 x 7, and additionally, each time you enter a new growth phase, or new wrung on the spiral of evolution, you the initiations start all over again. Thus, you do not go these initiations only once, but over and over, in different contexts, dimensional advancement, and phases of life; for life is a never-ending process of conscious evolution.

One life-transforming experience, mentioned in an earlier chapter, involved finding the unpublished manuscript written by

Charles Fillmore. In August 1947, just a few months before he made his transition, Mr. Fillmore hand-carried a manuscript to the Archives, at Unity headquarters. On the manuscript were written these words (paraphrased): *"Keep this until the world is ready."*

The manuscript was entitled, *The Story of Jesus' Soul Evolution.* In early spring of 2006, when I lifted the hidden manuscript from its resting place, the moment my fingers touched the paper, my hands began to vibrate. A palpable energetic thrill jolted throughout my body, like an electric shock. The vibration was mesmerizing and held me captive for several minutes.

In the midst of this vibratory experience, a feeling came over me, a sense of knowing, that this manuscript needed to be published, and that I was to make sure it happened. I realized that it was time to bring the manuscript out of the archives and place it in the hands of those who were ready for its teachings. For those who had *eyes to see,* as Mr. Fillmore states in its introduction.

The manuscript is a linear timeline of the events of that occurred in the life of the master teacher, Jesus, and as recorded in the four Gospels; Matthew, Mark, Luke, and John. Each character and element of the biographical story is metaphysically interpreted and represents an aspect of the soul's journey that will be met along the way.

Mr. Fillmore referred to this timeline of events and life-experiences as the Christ template, the Christ blueprint, for soul evolution. Energetically, for me, as I delved into the contents of the manuscript and consciously engaged with the ideas within, the vibrational level of my being shifted. The shift was a result of an initiation into a deep soul-cleansing process, which at the same time was lifting me into new and higher levels of dimensional awareness. Within six months of discovering the manuscript, a few friends and I had the original manuscript typed, paragraph for paragraph, and word for word. It was ready for distribution and publication; the time had come. Obviously, the world was ready!

Overtime, as read and reread the manuscript, I began to see clearly the Christ template and blueprint that Fillmore was referring to in the manuscript. I also saw that the Christ template contained step-by-step directions on how-to develop and establish the spiritual mind and spiritual mind discipline, along with other areas of soul growth required for evolution and ascension into higher dimensions of awareness. Additionally, I began to see how the initiations that Fillmore described in 1930, overlaid the template as a foundational underpinning; it was a hand-in-glove relationship.

Jesus was initiated into the mysteries of the Kingdom of Heavens (KOH), and as a mystic and master of energy he revealed the way to that mastery, through the creative process of regeneration. His life experiences, metaphysically interpreted, reveal a template for how-to bring all the forces of your mind, body, and soul under the auspices of higher-dimensional awareness, through understanding the hidden wisdom in your life experiences. Experiences are a test for learning your soul lessons. You just have to recognize that!

> *Life she is a harsh teacher,*
> *for she gives you the experience first, and then the lesson.*
> ~

The intention for sharing the metaphysical and mystical energy underlying the "Christ" template are to support you in:

- Recognizing what might be happening when you are in a phase of initiation;
- Understanding the elements of the template through the lens of Fillmore's metaphysical interpretation of, *The Story of Jesus' Soul Evolution*;
- Understanding the elements of the template through the lens of my more contemporary-oriented metaphysical interpretation, one that is steeped in the quantum language of energy, vibration, and frequency;
- Delving deeper into your own interior journey to unlock and unfold the mysteries of your own unique path soul evolution, and to do this through the exploration of metaphysics, mysticism, and energy dynamics supporting each initiation, element, and character found in Jesus' story of soul mastery;
- Releasing all vestiges of your religious past, and/or Jesus issues that might be still be holding you hostage and preventing movement into higher levels of awareness;
- Understanding the definition of regeneration and what it means to bring all of the forces of mind, body, and soul under the auspices of Christ Consciousness, 5th dimensional and higher awareness, and how that is accomplished through the choices you make, or choose not to make;
- Understanding, with a greater clarity, the work of the 19th and 20th century mystic, Charles Fillmore.

As you begin your sojourn into the discovery of the energetic dynamics and metaphysical-mystical meanings supporting the journey of the man named Jesus, and his unfolding of the Christ template (which is now available to all who are ready to say, yes); here are a few final considerations.

In each initiation, every element and character, is important to understanding the context of your soul's evolutionary process; and the experiences that you will face in life, on your souls journey. It's all there, stay awake, and you will see these characters and elements as a part of you.

When understood from a metaphysical and energy dynamics perspective, every character or element has the capacity to add layers of richness to your soul growth, through translating their experience into something that you have experienced. Also, note that the energies of the elements or characters, when introduced, often appear to be stuck, concerned, or just plain steeped in 3rd dimensional dense energy. Observe how they often must transform before the initiatory event can occur.

Then consider what this means for you life. How have you been a similar situation? When have you doubted, wanted to run, had to do something you didn't want to do? When have you faced your shadow-self, and was appalled. It's all in the template. Watch for it!

For example, in the birth initiation, one of the early characters introduced is, Joesph. When Joesph found out Mary was pregnant, he wanted to take her away and hide her. Then an angel (a thought messenger) came to him; and he had a change of heart. A new realization arose and it gave him hope and resolve; he knew he could do this! Think about a time when you were in the midst of change, or creating a whole new paradigm of reality around who you are.

Perhaps the experience came in the form of a new job, getting married or divorced, losing a loved one, a baby, an identity crisis, etc. Did you want to run away and just put it all behind you? Did you want to just stay in bed with the covers over your head? And then, a moment of clarity arose, however it happened, you had a change of heart. You decided to stay and face the next chapter of life!

Remember, you are not doing this alone, Universal Presence is here to support you in your process of discovery and possibility; in unraveling what each element and character means for your unique experience of soul growth and unfoldment.

Additionally, as life is an evolutionary spiral, you will experience all seven of the initiations, and the layers within them, over and over, as you rise up the spiral. It is not a once and done thing! With each rung of the evolutionary spiral, you will have another round of similar experiences. And each initiation has layers and layers of energetic dynamics underneath it, which is what each element and character comes to reveal! However, the good news is that with each turn of the spiral, the time it takes to recognize a dysfunctional pattern, and then make a conscious choice for something better, decreases.

Here is an evolutionary spiral scenario. You have a pattern of selecting partners who tend to treat you with a lack respect. Once you engage in the awakening process (birth), you begin to see this pattern—collected dense energy, shadow stuff, and you choose to start to consciously work on it, to transform it. As you transform, you realize, I deserve better, I am enough; and you make a choice to leave a ten-year relationship, and all of its dysfunctional patterning. Ahh, you climbed a rung on the evolutionary spiral. At the next rung, you meet someone new. But instead of ten years, you see the pattern in ten days, and you make a choice to move on, because you deserve better. Next rung, ten minutes. This goes on until there is no more dense energy, no more vibrational attractor in you, to draw those types of relationships to you.

For ease of understanding, each initiation is identified with a heading, and every character and element is identified with a banner. The banner states, ***The Energy Supporting***... It signifies that the character or element, holds a specific energy for soul growth; and the metaphysical-mystical interpretation of that character or element, can support you in seeing how this energy might be showing-up in you. The interpretations are either from Mr. Fillmore, or from my own personal experience; and they are only meant to be touchstones. The truth of what each character and element means, lives in you, unravel it!

What are the possibilities, gifts, or opportunities in this experience?
Napoleon Hill
(paraphrase)

1st *Initiation: Birth*

Birth, as an initiation, is an invitation to energetically awaken, to your inherent divinity, the truth of your being, the I AM. This first awakening is just an inkling in mind, of the possibility that you are something more than you thought you were. However, it is only a glimmer of your inherent divinity; for in order for the Christ idea or 5th dimensional reality (you choose the name that you are comfortable with, for this movement of energy) to be fully realized and embodied in you; you must evolve through the evolutionary spiral of spiritual transformation.

Mr. Fillmore says (a paraphrase), "More often than not, when people are first introduced to a Unity or New Thought service, a feeling comes over them. It is a sense of—*I am home*—and they begin to cry. That sense of home, is a sign that the inherent divinity, the I AM, has been sparked and It's desire to awaken is ignited." The birth initiation is commencing.

The Energy Supporting Inherent Divinity—in You!

Because the true nature of your of being, I AM, is hidden—you must unravel it—layer by layer. Yet, before this birthing process can begin, you must say, yes! You must on some unconscious level of your being, agree to initiate the process! As you say yes, you are initiating the journey into soul mastery and energy facilitation. Interestingly, some say yes to the journey for only a short while, because it takes too much effort to realign consciousness and make the changes required to transform.

The promise of inherent divinity is found in many sacred writings. Here are two passages, that when metaphysically interpreted, support the promise of inherent divinity. One is found in Genesis (Hebrew Testament) and the other in John (Christian Testament).

1. Genesis 1: 26-28, "*Let us make them in our image ... male and female, let us make them...and give them authority and dominion over all things, and the power to subdue all things.*"

Let us make them in our image, what is God's image? Webster Dictionary says that image is, "*a mental representation of anything not actually present to the senses.*" Mr. Fillmore, in *Revealing Word*, share; "*Everything manifested was first a mental image and brought into expression by the forming power of the imagination.*"

In this context then, Higher Power or Source Energy had a mental image picture of you as a divine, masculine-feminine balance, in a luminous light-body. However, because of freewill, you (and the collective) got to shape that image. Regardless of how you have chosen to cover up your divinity. It doesn't change the idea that underneath all hidden layers of density, of skin, bones, cells, beliefs, perceptions, thoughts, and emotions; you are inherently divine.

The promise is that, at level of structure, you (all) are an energy field, a vibrational-light-being created from source energy. With that promise, there is also inherently conferred upon you, the capacity for dominion and authority over all things (all your thoughts and feelings). Meaning, you are not a victim, you have dominion and authority. If you choose to claim it, align with it, and use it to make conscious choices! You must say, yes, to it!

Next is the power to subdue. You have, energetically, the power to subdue (overcome, quiet, or conquer) any thought or feeling that is not in alignment with the higher-dimensional awareness of the I AM, your true Self. This is not about having dominion, authority, or power over others, but over any limiting belief, thought, or emotion that you have. Wow! You are awesome.

2. John 1: 1, "*In the beginning was the Word (I AM) ...and the Word was God...and the Word (I AM) became flesh and dwelt among (in) us.*"

Word/I AM is and holds the true power, authority, and dominion. When you claim I AM/Spirit as the dominant vibration of your being, as the only power and energy working through you—you activate your capacity and power to subdue, overcome, all 3rd dimensional adverse energies trying to hold you hostage to their density; all that does not allow the light that you are, to shine through!

The Energy Supporting Lineage and Prophecy

Lineage and prophecy are precursors to the birth, and are important for they refer to the early, early, characters and elements that supported the preparation of your soul and mind for receptivity to the birth initiation. Lineage represents the past, and the laws, qualities, perceptions, and beliefs that were "indoctrinated" or shared with you as a part of your family values, cultural beliefs, and societal norms. All of these shaped what you currently believe about your culture, tradition, religion, family history, family illnesses, and more.

They are a part of your embedded "linage and heritage" that ultimately needs to be transformed, before you can engage the fullness of the spiritual journey.

Prophecy is also a part of your past. What does your family lineage hold as a prophecy for who they believe they are? What does the family believe they will experience, because grandpa had it? What is in their belief system that defines their future, and yours? Were you a wanted or unwanted pregnancy? Where did you come in the family line-up; first-born, middle child, only child, or baby?

Prophecy also includes such things as, your genetic make-up and what affects your family from a physical and mental health standpoint. This includes diabetes, heart disease, alcoholism, weight, cancer, etc. Each of these presents its own prophecy for your life, if you accept it, and ultimately, as you will see, each must be lifted up into inherent truth of who you are. And that truth is that you are the expression of Spirit, Source Energy, One Mind, I AM, and that is your only and only true lineage and prophecy.

The Energy Supporting Zachariah and Elizabeth

Genesis 1: 26 and JN: 1, set the stage as the promise of your inherent divinity, and thus, your innate capacity to return to that state, for soul evolution. Zachariah and Elizabeth energetically represent the first movement of energy prior to the birth initiation.

As a couple, they represent a level of promise for soul evolution. They have an insight into God, but they are still spiritually immature. They are barren of child, meaning even though they are spiritually devoted they have not brought forth the true fruits of spiritual maturity. They are still spiritually immature, living from the intellect alone.

Zachariah and Elizabeth are devoted and obedient to God, but their I AM awareness has not been awakened or stimulated, for their embedded dogmatic, religious thoughts, hold them hostage to lower dimensional energies. These thoughts must be lifted into a higher state of awareness for the fruits of Spirit—higher-dimensional awareness— as a reality to be realized, made real, in their lives. Do you remember when you were asleep to your divinity? Maybe you still are?

Thus, their pregnancy, is a miracle of sorts. It results in the birth of John (the Baptist). John is the precursor to the Christ but is _not_ the Christ. All elements must come into play in the right way, and in the right time, in order to birth a glimmer of Christ awareness.

For me, I was spiritual asleep, living on the east coast, and my husband got transferred to Kansas City, Missouri. I had to agree to go with him, my yes. Three days in K.C., and I met a woman who told me about a spiritual center she attended, and invited me to go, it was Unity! It was a miracle, of sorts, I was home!

The Energy Supporting Mary, the Divine Feminine, and the Annunciation

This aspect of the journey takes place in Nazareth, which energetically represents nothing special, a commonplace in mind. This metaphysically means that you do not have to seek for it, go a mountain top, or do anything special for the first glimmer of awakening to occur. You don't find it -- it finds you!

The angel Gabriel, a representative of higher dimensional awareness and thought, makes the announcement to Mary. The annunciation is a pronouncement or announcement that something is about to happen in consciousness – get ready–you will need to make a conscious choice about it, when it happens.

Remembering, involution and evolution, everything starts with the involution of an idea in consciousness, then evolves outward from there. So, for the announcement to have any effect on Mary, she has to agree. She has to say, "yes" to the receiving of the gift of the idea of Christ awareness. To say yes to soul evolution and takes surrender, humility, and the capacity to establish the feminine qualities of being in mind.

Thus, Mary represents a prepared soul. A prepared soul is one who is ready to initiate a new phase of understanding in mind—albeit it is an unconscious awareness—regarding the feminine nature of divine.

The feminine nature of the divine resides within each person, regardless of whether they are in expression as a male or female body. Please note that each character portrayed as a woman in the template, represents an aspect of the feminine nature, in various stages of development moving towards the full restorative and regenerated power of the feminine nature. The same goes for the male characters.

You cannot birth Christ consciousness without the influence of feminine energy. It is regenerated feminine energy that calls forth higher levels of unconditioned love, compassion, harmonic resonance, and balance within being. This is important for the experience of sacred union, at Cana, can only occur when each aspect of the soul, masculine and feminine, has grown into their own unique wisdom and authority; and is now able to stand in balance and union with its counterpart—without any need to be dominant!

In my book, *Embracing the Feminine Nature of the Divine*[36], are quotes from Charles Fillmore, on the metaphysical importance of Mary, the mother of Jesus. Fillmore purports that Mary is the first representative of the feminine nature of the Divine, to appear. Her appearance heralds the capacity for a spiritual birth to place:

"The conception and birth of Jesus, as recorded by Luke, conceals and to the spiritually wise, reveals a soul principle that will save man from death. That Principle represented by Mary [his mother] is Love. Up to the time of Jesus the feminine principle of the soul, love, never had a chance to express itself because of the arrogant dominance of the intellect.

Jesus would have utterly failed in the resurrection of His body after the Crucifixion, if He had not developed the restorative power of Divine love. So, no man can hope to escape death until he frees the imprisoned love of his soul."

"The [intellect] masculine phase of mind has been allowed to dominate for so long that it now assumes that everything must be subservient to it and that its dominion and dictation is the edict of divine law. The [love-intuitive] feminine phase of mind, with its mighty heritage of love, has been mesmerized into this belief [of intellectual authority] and accepts as a matter of course the rule of head over the heart.

This false state of mind has thrown the whole race out of balance and severed the spiritual connecting link between man and God, which is love... Jesus Christ was a balanced combination of wisdom and love, masculine and feminine."

Until you consciously awaken and understand the role the feminine nature of the Divine plays in regeneration and soul evolution – whether you are male or female in bodily form – you cannot fully proceed in your spiritual journey into the full realization of 5th dimensional awareness, Christ-consciousness, Buddha nature, etc. The Christ nature is "begotten" from love (Mary/feminine qualities) wedding wisdom (Joseph/masculine qualities); and compassionate Christ awareness is the offspring of this sacred union.

---

[36] Toni G. Boehm, *Embracing the Feminine Nature of the Divine*, Inner Vision Press

Saying, yes, to the spiritual growth being offered at my new-found spiritual center, was the first spark of the feminine nature of the divine, love, coming alive in me. Mary in me, said yes! I just had to grow into what we had said yes to!

The Energy Supporting Holy Spirit Overshadowing

After Mary says "yes", the *Immaculate Conception* is brought about through the energy and power of the universal impulse, the Holy Spirit, "overshadowing" Mary. This new energy awakens a new level of "living" wisdom, a first glimmer of the illumination of the light of truth, and the notion that there might be something "bigger" at work in the universe and in your life.

Know that before the heart can conceive the Christ idea, experience a spiritual unfoldment, or an awakening there must be a movement of higher dimensional energy, which comes in the form of light-energy. *"The Holy Spirit shall come upon thee and the power of the Highest shall overshadow thee," "The Light that lights every person comes into the world."* This is the beginning of your transformational experiences.

The Energy Supporting Joseph

Joseph metaphysically means, *outer mind practicing obedience to the Lord, but not having understanding.* Joseph energetically represents the non-understanding aspect of masculine energy. He was of a mind to put Mary away, to banish her, for he did not understand what was happening. He wanted it to go away, but a higher thought prevailed. Have you every experienced a time, in the midst of your soul growth, that you just wanted it to be over, you cried out, stop, this is too much? Then you paused, took a breath, and became willing again. That was a Joesph moment.

The Energy Supporting Mary Visiting Elizabeth

Mary (feminine phase of love on a spiritual plane of awareness) goes to see Elizabeth (intellectual plane starting spiritual growth). They are drawn to each other, and in that drawing together Elizabeth (steeped in the level of intellect) starts to recognize the importance of spiritual growth. She receives a sign of validity of her new thought, her baby jumps in her womb. Do you ever have a sign of the spiritual validity of things? Like, God bumps or chills?

The Energy Supporting Gestation

After conception, there must be a time of gestation–of digesting and allowing an internal growth of your new ideas. Gestation refers to the time needed for prayer, attending classes, reading spiritually-oriented books, and studying to establish the seeds of awareness that are beginning to grow. Seeds, or divine ideas, need an established root system or they can easily be destroyed by the first adversarial wind that arises.

The Energy Supporting Going to Bethlehem

During the gestation period, you must go to Bethlehem in order to pay taxes and to partake in a census. Bethlehem means place of substance, and metaphysically, substance is the divine energy that underlies all things. The seeds of truth you are planting need spiritual substance to grow, before they can be birthed. These new seed ideas rely on "deepening-type" activities to establish them in consciousness. Activities such as self-observation, being open to learning, prayer, forgiveness, self-forgiveness, use of denials and affirmations, etc.

The Energy Supporting Census and Paying Taxes

The energy behind the census is about being counted, being accountable, being recognized as part of a "tribe," (a spiritual tribe) and being willing to be counted as one of its members. You realize that you count and are accountable for what you believe. You realize that you are accountable for your actions — for the words you speak or do not speak, for the up-holding of your new beliefs, and to "leave" old habits and ways behind. As you realize all of this, often this initiates a change in your life. If you have any karmic debts (pay taxes) you need to start to reconcile them. When you pay taxes, you are beginning to pay off karmic debts, paying your dues, creating a foundation for new thoughts and thought substance.

The Energy Supporting No Room at the Inn and Not Giving Birth in Public

No room at the inn metaphysically means that often in life, you are so occupied with the things of life and outer activities that there is no room for spiritual ideas to be birthed. Sometimes it takes a humbling experience for you to remember, what is important.

Also, birth is private, it occurs interiorly in a prepared space. Albeit, a humble place. It takes time to prepare that place, and a commitment.

The Energy Supporting the Manger

Humility is a necessary attitude for spiritual growth and unfoldment. Thus, you give birth in a manger, signifying a "lowly," humble place prepared in mind.

The Energy Supporting Giving Birth at Night

The idea/child is born in mind, in the stillness or silence of night, when intellect is quiet and when you are closest to instincts of intuition. Remember, you had to go through all these other elements, and you are just now reaching the point where the glimmer of divinity, is about to fanned into a flame.You are about to give birth to a whole new level of awareness, and truth about who you are!

The Energy Supporting the Beasts

The birth takes place among the beasts (the thoughts of the unregenerate or unhealed mind). Spiritual illumination will grow and dispel the darkness of unregenerate mind, overtime. But you will need to consciously do the work required.

The Energy Supporting Being Wrapped in Swaddling Clothes

The newly birthed Christ idea is wrapped within the physical limitations of the body and therefore, it will need to overcome the dense energies, that are still hiding it. Therefore, you must take on the task of growing this new level of awareness, into the fullness of its spiritual stature.

The Energy Supporting the Star in "East"

The star from the East is an interior light that comes from within, from the birth of Source Energy. This light is a greater light than intellect alone, and yet, it is but still just a glimmer of the full light of the "Christ" Idea.

The Energy Supporting Shepherds Watching at Night

Being vigilant over your thoughts and feelings (sheep), is important in order that you protect what has just been birthed, and so that you remain receptive, obedient, and spiritually disciplined—in those moments when you become fearful of the change that is happening to you. *The shepherds tremble and are afraid; "...when new vibratory forces start to rise, there is a trembling in mortal parts."* Charles Fillmore

The Energy Supporting the Three Wisemen/Kings

These three stored-up resources of the soul rise to the surface when the soul's depths are stirred by a great spiritual revelation, and/or by an ascension in conscious to higher level of awareness. The inner realms of consciousness have sacredly held these resources, the records of the wisdom gleaned from past lives, in reserve until the great day would arrive when the soul would receive the I AM.

The Energy Supporting Herod

Herod is your adverse ego, and it still rules the un-regenerate thoughts that are hanging out in your mind. Thus, you must commit to being devoted to protecting the "child" from Herod thoughts. Herod represents and reveals the hostility and jealousy that is still present in you, and that is held in human or carnal mind. Herod desires to kill-off the Christ child.

e.g., Don't go to class today, you know so much already, in fact, you are probably just as smart as the teacher. What harm would there be in hanging with my old friends?

The Energy Supporting the Three Wise Men or Kings

Three Wise Men or Kings represent the inner resources you are now developing and that you now have at your disposal. They have been stirred and called forth by all the new revelations you are gleaning. These inner resources include; gold (riches of Spirit), frankincense (beauty of Spirit), and myrrh (eternality of Spirit) and they are available for you to use in support of growing your new Christ Idea.

The Energy Supporting Simeon

The Christ child is now in your conscious care, and you must be committed to its care. When Simeon blesses the baby, the power of that energy of blessing represents, an impartation of energy and the quickening and multiplying power of Spirit that produces growth and increase in mind. Blessing represents the power of multiplication —which is the secret key to abundance and prosperity. You receive this energetic blessing as you align in consciousness with the higher vibrations of gratitude, acceptance, trust, compassion, and more. This is part of staying focused on what is truly important!

The Energy Supporting Circumcision

At eight days the first attempt to come close to God is demonstrated through the symbolic act of circumcision. Circumcision is an outer ritual that is a symbol of the covenant between God and the Jewish people. It is a symbol of the intellect's first attempt to come close to God, by the cutting off mortal tendencies. It is the first act of redemption you consciously partake of, on the path of "salvation," of redeeming your thoughts. It might be a thought of forgiveness of self, of others, or a sudden realization of understanding who and what you are in truth. It might be someone asking you to do something you used to do, and no longer do for you have outgrown it. Saying, no, is act of redemption.

The Energy of Salvation & Redemption

Redeeming your thoughts, means to bringing them back under the guidance of Spirit. Redemption is the obtaining of spiritual consciousness on "earth" and is represented by the physical body. The body is the focal point through which the I AM functions. You are "saved" or receive "salvation" as you energetically redeem your thoughts, as you raise them to new heights. Sometimes, this means that you have to no to old habits or friends. Things that are trying to sneak-up and gain a new foot-hold.

The Energy Supporting Joseph & His Dream

Your conscious mind is beginning to recognize the energetic importance of dreams and visions. You begin to see them as guiding forces being sent by your Higher-Self. Thus, you protect, and keep safe, the energy of your new ideas as you are guided by interior wisdom. A great spiritual practice is to start journaling these dreams, so you can watch your progress. I did this for years. I love to go back and see how my dreams were revealing next steps; providing answers to questions that hadn't even arisen, yet; and how often, they were precognitive, showing me what was coming.

The Energy Supporting the Descent into Egypt

At some point you will go energetically down into Egypt. The energy of your spiritual wisdom is in its infancy and it must be protected from the energy of Herod's 3rd dimensional unregenerate thoughts. The killing of the first-born children represents the elimination of the unregenerate thoughts in order that new energy, a life-giving thought energy can be established.

This is the conscious work that must done and it often involves making major life-style changes, hard decisions, and transforming unregenerate thoughts of not enough, into I am awesome!

The Energy Supporting Herod's Death

When Herod dies, you leave Egypt, your place of energetic safety, and you return to Nazareth. You have been preparing yourself for "twelve years," and are ready to energetically stretch-forth. You are now ready for an even greater influx of spiritual energy, wisdom, and development, and all the experiences and lessons that that entails. You are ready to be about sharing what you have learned. Sometimes this first step is scary. Mine was! I was asked to teach a prosperity class. I told my minister, no way, not me, I don't know enough! Yet, never-the-less I decided to be willing – to stretch my wings and fly – and I did it! It was a spiritually stretching experience.

The Energy Supporting the 12-Year-Old Mystic

Remember, energetically, however, you are just a teenager in your soul growth. You are growing in wisdom and stature, and people are recognizing that energetically, your vibrational field is changing. You are clear that you are ready to "be about your Father's business"– you are energetically prepared to move forward into new levels of dimensional awareness—spiritual business.

You are ready to leave the energy of your "old" life behind, and pursuit ideas, experiences, energies that more in alignment with who you now are. This may mean a move, a commitment to greater involvement in your spiritual community, and more. And remember, at this point you are teenager, you often think you know more than you do, stay open to learning!

The Energy Supporting 18 More Years of Spiritual Maturation

Twelve and eighteen add up to thirty. Energetically, it takes *"thirty years"* to establish the qualities in mind that support the journey of spiritual maturation. Thirty years brings you to the vibrational and spiritual readiness that heralds in the energy and experience of the first baptism. This is the point, where you will be required to step into deeper, more honest overcoming and transformation. You will have to initiate a journey into consciously transforming what you see as your shadow-stuff.

This is NOT an overnight journey! It requires patience, commitment, and spiritual mind discipline to evolve into this energy field, and at any step along the path, you can say, *"I give up, this is energetically too much for me."* Many do!

The Energy Supporting Eighteen Years and the Three Vipers

Mr. Fillmore shares that people are always asking where did Jesus go to study during the eighteen-year period, when nothing was written about him? He eludes to the fact that perhaps, that isn't the right question. He says it is not about where he went, but what did he do? What did he work on?

Mr. Fillmore says that during this time there are three vipers you must energetically transform, that you must overcome, and you need an eighteen-year period to do it. So it isn't a fast or quick fix. Vipers, are those lower 3rd dimensional energies such as, jealousy, greed, covetousness, family and religious influences, cultural heritage, need to control, me-me-me, etc. All of these, and more, must be energetically transformed, overcome, in order to be vibrationally prepared for the next initiation. Remember, eighteen years, means, as long as it takes. The three vipers you must energetically work on are:

Tradition & Heritage

Selfishness

Abuse of Power

1. Tradition and Heritage

Many dense and negative-oriented energies, in the form of beliefs, perceptions, and tendencies lurk within tradition and heritage. All of which must be overcome, for they hold you energetically, as a hostage to their lower vibrations. These energies revolve around your lineage, your family beliefs, culture, family heritage, importance of family, place and role in family dynamics, religious upbringing and traditions, race, gender, what your mother or family did to you, etc. The work is to into a mindset of One Energy and Power—Spirit—Source Energy—only!

If you have not embraced, overcome, and healed your family dynamics and/or religion or origin issues, you cannot climb the spiral of evolution, you will be stuck in 3rd dimensional energy. You have to embrace to erase.

2. Selfishness

You must learn that is not always about me, me, me, and did I say me! My wants, my desires! You learn to share your good. Selfishness includes the energies of hypocrisy, pettiness, condemnation, antagonism, vanity, use of platitudes, greed, and more. You must learn to be less concerned with the material and with material things–not that you cannot have them— but you are no longer energetically attached to them as status symbols, or yielders of power.

Getting over selfishness, orientation towards me, me, me which is a step in the process of the individuation—for moving from victim to individuation usually involves finding the inner strength to be OK, with it being about me, me, me. However, there comes a point where, in order to move into true sacred service, you have to get over yourself. And that usually involves an identity crisis, of some sort.

3. Abuse of Power

There are two types of 3rd dimensional energies that support the abuse of power. The first-type of abuse of power occurs when you play too big. You unconsciously and energetically sense, I am not enough, so you take on a façade and pretend to be something you are not.

You engage in activities and behaviors, such as; trying to control everyone, situational control, acting powerful, authoritarian, manipulation, bossy, being overly strict, getting your way, taking what isn't yours, being greedy, having to seen as right all the time, showing how smart your are, thinking you know what is best for everyone, being overly attached to money, success, power, making others feel small so you can feel big, engaging righteous judgment to criticize others, protecting turf, and more. You use those energies to energetically feel better about yourself or to get your way; but in the end they hurt people.

The second type of abuse of power, is one that you would not often think of, and that is being so afraid of who you are that you play small. Playing small is an energetic abuse of power, for you engage it in order that you can; play victim, poor me, be seen, or get attention.

Both types of abuse, however, arise from a 3rd dimensional level of energy that holds a lack of authenticity and transparency. Thus, at some point on the journey you will need to learn how to engage in the energy of honest self-observation, transparency of thought and emotion, and standing in your authentic power. You must release the energy behind the belief that any of these power-oriented behaviors can bring you importance or happiness. They don't, they are a part of the energy of your shadow self that lives hidden in consciousness. And again, to release them, may include experiencing a major identity crisis, or loss of something you deem important to you.

<u>2nd Initiation: The Two Baptisms—Water & Spirit</u>

Using Jesus' life as a template for your spiritual journey, thus far, you have energetically moved through your infancy, childhood, teenage, and young adult years, you are maturing. You are growing in spiritual stature and maturity and are entering early spiritual "adulthood." However, you are still young in your learning; there are many energies, experiences, and lessons to uncover and discover. It has taken thirty years of renewing of the mind, and you are still evolving into a greater awareness of who you are. However, at this juncture, the intellect is still leading.

The Energy Supporting the 1st Baptism—Water

The baptism of Jesus by John energetically represents "sense-man," 3rd dimensional being. Having been cleansed through the "pouring of the word" upon the intellect and purging the mind of the energy of error thought; this work has been accomplished through the conscious use of denials, affirmations, changing of the mind, forgiveness, meditation, and more.

The intellect is becoming illumined (the John the Baptist phase), but it is still only a first stage in the development of higher vibrational awareness. The illumined mind is not the intuitive mind of higher 4th dimensional awareness, it is a precursor to it. There are many more levels of multi-dimensional awareness to move through.

However, because you appear to be a spiritually bright light, and perhaps, even think you are; you give great meaning to your new sense of illumination. The adverse ego, survival-oriented ego-personality, thinks it is spiritually special because it is showing such great leaps in soul growth. You might even be doing wonder-filled things like energy-work, energy-healing, speaking to crowds, teaching classes, etc. You might believe that these activities make you special, or give you the idea that, energetically, you are way ahead of others.

Often at this stage you start to believe your own hype, about how special you are, beware! This is very a subtle stage of awareness, for the intellectual ego is very tricky. I learned a very valuable lesson, during this stage, which only later, was I able to see.

I was a newly ordained minister, teaching prayer, to a large, in numbers, class. In the group was an elderly gentleman, in his ninety's. He had been on the spiritual path for a long while and added much to the class experience. After a meditation that I had led, he came up and shared with me that it was the best meditation he had ever experienced. I felt myself preen and begin to fill up with pride. Wow wasn't I especially great!

In my sense of swelled-up pride, I asked him which part he liked best. Was it my voice, my words? Just what was it, that I was so good at delivering? He looked at me and said, "oh honey, I can't tell you that, I didn't hear a word you said. I am deaf!" It was like someone hit me upside the head with a 2x4.

In that moment, I was humiliated (ego) and then I was humbled (Spirit). I had just learned a great soul lesson. Whatever happens in the classroom, is not of my doing; it is Spirit at work. Take credit for nothing. Soul lessons are not about good or bad, but about learning that every experience is a part of your transformational process.

You must become fully prepared for the journey to the interior depths of the heart, and sub-conscious mind. For here, there is deeper spiritual work to be accomplished, and at times it can appear touch, or even ugly.

Your intellect is unfolding in wisdom, but you are still steeped in intellectual knowing and intellectual based-ideology. You are repenting, energetically shifting and transforming old beliefs and perceptions, as you are being vibrationally prepared for the sacred initiation of "fire" by the Universal Impulse-Holy Spirit. The energy of the "Christ" consciousness is slowly being revealed, and yet it takes a dominant vibration that holds the energy of knowing who and what you are, as the I Am in expression, to continue on this journey of regeneration.

The spiritual path is becoming noticeably narrower and straighter, meaning that the ideas, words, energies, vibrations, frequencies, and actions that you engage are rebounding quicker. The Christ vibration has been in your care and you have been faithful in guarding it through a conscious choice of your words, thoughts, emotions, and actions. You have been committed and dedicated to its growth.

You have taught the "doctors and rabbi's," the learned thoughts of the intellect, as you know that they too "must learn to listen to the Christ Mind." You continue to persevere in your learning, working consciously to change through gaining more knowledge about who you are and how you are "showing-up" in life.

You are mastering the energy of self-observation, being faithful in your commitment to prayer and meditation, and in being less reactive to people and experiences. You are attending classes, finding outer teachers, or whatever you perceive might work and bring you into a higher-level dimensional awareness.

Still, you are trying to spiritualize the intellect through more information, and it doesn't work! The intellect does not readily receive the Christ vibration. The unregenerate thoughts of the intellect doubt and question the validity of the Christ vibration, and the new internal reality it is working to establish. The intellect, with its connection to then adverse ego, would rather try to make the Christ vibration and message fit within their parameters, rather than shifting into a new level of awareness. e.g., you might find yourself saying:

Maybe I can grow in awareness and still hang out with my old drinking buddies or at my old hangouts. Maybe I can raise my vibration and still "steal" a few dollars from my mom or job. Maybe I can grow in awareness and still have an affair, after all we are on the same wavelength – and my spouse isn't. A little white lie here or there can't possibly hurt.

You continue to express your spiritual ideas in intellectual terms; you intellectualize your spiritual concepts, and often try to show others how knowledgeable you are of spiritual things. The intellect, although raising in vibration and gaining new awareness and knowledge is still in darkness regarding the things of higher dimensional awareness. Also, you are just starting to realize that soul mastery is an absorption process, not an intellectual process.

The energy held by the intellect, or illumined intellect, is not wise, nor does it necessarily hold great wisdom. It is smart, and full of information. It has done great clearing work. It is this clearing of old energy, dense energy, that allows enough light through, to be considered, illumined—and it is still not enough. For although the illumined intellect has been spiritually prepared, it still holds a great deal of doubts and fears.

You have been voluntarily cleansing your energy field by rejecting and releasing "sin" and "error" thought and by consciously raising your vibration through the use of denial and affirmations, forgiveness, prayer, meditation, etc. However, to anchor this new vibration in mind, you must initiate a full surrender into unlearning everything you thought you knew. The mind and inner heart are being brought energetically into coherence and being prepared for the great demonstration that is coming; a time when you will be asked to demonstrate what you have been energetically absorbing and vibrationally anchoring.

The completion phase of your intellectual preparation is metaphysically represented in scripture, when John is beheaded. Metaphysically, this means that you have cut off of remaining vestiges of the hold of the unregenerate, 3rd dimensional-oriented intellect, even the illumined intellect. To begin the descent/ascent of into higher levels of "heart" wisdom.

The first baptism is that of water, and the second baptism is that of Spirit or fire. Water cleanses, but fire scorches! Get ready!

The Energy Supporting the 2nd Baptism & Quickening by the Holy Spirit

The intellect must lose its hold on the mind, in order to receive the quickening vibration and fire of the Holy Spirit – Divine Love. Divine Love is in a more contemporary interpretation is, a harmonic field of resonance that is unconditioned by past beliefs. The mind and being are ready, they have been renewed, transformed, and regenerated on many levels. As you *"transform yourself, you transform the world." "Be ye transformed by the renewing of your mind."* Romans 12:2

Harmonic resonance is not only a field of energy that is unconditioned, but it is also, a balance of opposites and polarities. This balance, harmonic resonance, is metaphysically revealed and referred to via relationships. The relationships such as; Zachariah/Elizabeth, Mary/Joseph, Jesus/Magdalene, head/heart, love/wisdom, subconscious/conscious minds, water/wine, and more.

The second baptism, the Baptism of the Holy Spirit, lights Its fire at the center of your being and raises the vibration of your soul and body to high degrees of purity. Prepared rightly, you are now ready to cleanse the deep recesses of the subconscious mind. You are prepared to energetically transform your shadow aspects/energetic density, karmic or otherwise. You begin this process by washing in spiritual thought, represented by going into the river Jordan. Jesus went *into* the Jordan river *to be* baptized.

The dove coming down at the moment of the baptism represents the Holy Spirit's quickening power of love, represented by the feminine nature of the divine. It is the feminine nature and its power of intuition that ultimately reveals the Inner Voice. A voice that is recognized as different from your constant mind-chatter of the intellect. The Voice is recognized, as that of your inner spirit; it is heard only by the one receiving the baptism of light. It shares the sacred message of — *This is my beloved child, in whom I AM well-pleased.* It means, it is time you for to be in service to the energy, and wisdom of the I AM, the fire and heart of Spirit. It is time for the I AM to be the dominant vibration in being.

Mr. Fillmore in 1930, equated the dove with the Holy Spirit, and the Holy Spirit with the Holy Mother, and with the feminine nature of the Divine, the Divine Feminine. This would mean then that the Holy Spirit baptism is an initiation that sparks a fire of awakening and possibility. The possibility to go deeper into a greater understanding of the feminine qualities of the Divine in you, as you.

The feminine represents qualities such as intuition, your intuitive nature, love, compassion, sacred service, humility, collaboration, trust, authenticity, transparency, clarity, the unknown, spiritual mystery, "sacred-luminous-darkness," womb, and more. You will need these qualities, for what is to come, after the baptism. Here are a few ideas on the role of the feminine nature of the divine.

The following are excerpts are from my book, *Embracing the Feminine Nature of the Divine;* they came from unpublished documents written by Mr. Fillmore. They focus on the metaphysics of the Holy Spirit and the second Baptism.

"I believe that the Holy Spirit, the Holy Mother
[the Divine feminine], is being restored to Her rightful place with
Jehovah [the Divine masculine], in the heavens and the earth."
Unity Tenet #16, written in the early 1930s

"In an analysis of the Holy Spirit, we find that it has been used by religious people in every age, but not fully understood ... The love

of God is [the] Holy Spirit... The Holy Spirit, let us say, is the Divine Feminine of God. We know that God is Wisdom, God is Love. Love is the Holy Spirit, or the Divine Feminine and that is the Holy Mother."

"Jesus Christ reached the limit of his masculine consciousness, he was the Light of the world, the heat of the world, but he could not express on that plane until He became the Holy Mother.... We must enter into that Holy Mother consciousness within us, we have not that concept of the Mother."

It appears that Mr. Fillmore believed that in order for Jesus to evolve into the fullness of his ministry, Jesus had to be baptized into a higher level of awareness, understanding, and connection with the Holy Mother, the Holy Spirit, higher dimensional qualities of his feminine aspect. What about this might be important, to his soul growth? I have come to discover, in my own journey, that is only through a higher heart-level connection and understanding of the feminine nature that you open the way for a deeper connection with your intuitive self. This creates a new vibrational level of awareness and capacity for the demonstration of compassion for humankind, the sacredness of being human, and for self.

Additionally, it is only after the feminine is regenerated and restored, that the sacred union of the masculine and feminine can occur—refer to the wedding at Cana. The Christ vibration is a field of harmonic resonance, a field of energy that is a balance of both qualities. Remember, feminine energy started your spiritual journey, it is the feminine that said, yes (Mary). Both the masculine and feminine aspects or powers within you, and the characters representing them, is an evolving aspect of metaphysical movement towards sacred union. The Law of Life is the generating principle of two, working in sacred union, as One.

My Holy Spirit baptism occurred at my mothers celebration of life ceremony. I was officiating for the first time, and as I stood up to walk over to the podium, I knew I was going to faint. I couldn't do this! In that same instance of doubt and feeling faint, a brilliant, blazing white light jumped from my mothers' casket, came up in front of me, and then merged with me. It ascended-descended up and down my body, undulating waves of energy, set me aflame! The heat was overwhelming.

In the midst of the brilliant, blazing light, a voice began to speak to me. It was my mother, she said to me, *"Toni in my life, I gave you birth, in my death, I give rebirth."* As fast as it started, it was over; but the essence remained. I looked around and realized that no else, had seen or heard anything, and I knew, that I knew, that I had just been baptized by the Holy Spirit.

Within 24 hours, I was catapulted into my first conscious wilderness experience, and my journey with the Divine Feminine was initiated.

3rd Initiation: Wilderness with Temptations

In the 1st Baptism, the illumined intellect consciously pursues what it thinks it needs to spiritually "grow." The 2nd Baptism is a tremendous movement in spiritual advancement, for through the activity of the fire of the Holy Spirit, you awaken a deep and abiding faith. It is this feminine quality of faith that will support you in the next phase of the journey. For next Spirit leads you into the deep recesses of your shadow-self. An unraveling process.

Remember, you are led into the wilderness by Spirit – *NOT* — the devil. You will meet the devil later on, and you will discover that the devil is not who think, s/he is.

The journey into the wilderness, begins with the experience of forty days of fasting and then the three temptations. Each comes in order that you might demonstrate how much you have really assimilated in consciousness thus far. For what you have learned, you must demonstrate! If you cannot demonstrate it, it is of little or no value – it is just another intellectual learning. So, the 3rd initiation, Wilderness and Temptations, is about demonstrating what you have spiritually learned, up-to-now. Even thought it may not look or feel like that!

In the wilderness, you will be led through a series of experiences selected for you, by Spirit. These wilderness experiences will reveal the depth of your spiritual growth, and they will not necessarily be experiences that you would have chosen for yourself, to work your growth through. Nor, will they necessarily be pleasant, or pretty.

Spirit knows how much you have really absorbed and assimilated in consciousness. You may be able to fool yourself, or others, regarding how spiritually advanced you are, but you can't fool Spirit. In the mind's wilderness, you are called to demonstrate, to set your demons-straight, and to know who and what authority is in charge of your thoughts.

You have to learn to discern energetically between what is truth and what is illusion or false, between substance and shadow. The fasting and overcoming of the three temptations of the wilderness experience supports you in proving you know how to discern the difference.

The wilderness and its experiences further develop you and awaken higher levels of spiritual discernment, intuitive guidance, and establish a greater awareness of the I AM as the dominant vibration of being. So, wilderness is a time testing and demonstration of spiritual principle. What Spirit reveals must be demonstrated through experience. There is no soul growth without demonstration.

Mr. Fillmore says that, *demonstration is about getting our "demons" straight.* Fasting, holding true to spiritual principle in the face of adversity, takes you to higher levels of vibrational awareness where in-harmonies, negative energies, and shadow-stuff— your demons—dissolve into nothingness. Thus, fasting is an abstaining from feeding on the energy of 3rd dimensional sense consciousness ideas and reactions.

Fasting for 40 days begins to increase your capacity for spiritual discernment. Forty is 4 x 10. Four metaphysically means foundation. Ten contains all numbers, thus we have completed building the first level of introduction into higher dimensional awareness. You now have what you need to move forward into the next phase of your spiritual growth. Because you have adhered to spiritual principle in order to call forth greater spiritual illumination and discrimination.

A secret, you want to know how spiritually developed you are? Your spiritual progress is determined by your level of non-reactivity. Your non-reaction to persons, things, and situations, this is not to be taken lightly. You want to know how much you have grown spiritually, watch your reactions.

The Energy Supporting Why Testing Through Temptation is Good!

Testing through temptation comes as a means of demonstrating the unfolding of the spiritual truth and power being revealed through you. You are proving what you are made of — Spirit.

"He that overcomes shall inherit all things." (Rev. 21:17).

Temptation comes for you to learn non-judgement and non-attachment, to spiritually discern if you (or the temptation) are in alignment with principle; and then make a conscious choice to release, if not. You must make a conscious choice to stand on principle. Mary Baker Eddy said, *"Through science or suffering we learn."*

Make no mistake, if your choice is to be on the spiritual path, you will have experiences that will invite you to overcome the "devil" in you, the adverse ego, the small self; with all its negative, adverse, and error thoughts and emotions. You will be invited to learn from their coming. Remember, that old adage, *"Experience, she is a hard teacher, for she gives you the test first and the lesson afterward."*

The Energy Supporting the Temptations

You will experience temptations and often these experiences will often feel like dark nights of the senses or of the soul, yet they only come to teach you about you, about yielding and surrendering. Joel Goldsmith says, *"Non-healing is a lack of yielding."* The wilderness is about transforming the adverse-ego, the little self; self with a little *s*.

You stand alone in temptation! The devil is the adverse ego, the personality of little self; is an illusion created by thoughts steeped in 3rd dimensional awareness, sense mind-consciousness. The adverse ego personality attaches itself to the sense perceptions, beliefs, and emotional residue created from the experiences of your life. It then uses them as reasons to stay "stuck" and entrenched in having my own way and not feeling safe.

The wilderness initiation supports you in redeeming the shadow-aspects that lurk in your subconscious mind. Shadows are energetic blocks that keep the Light of who you authentically are from shining brightly. It is your shadows that keep you from being transparent and authentic.

You are on the journey to transform your shadows. Thus, you must learn to be willing to explore your shadows, for eventually they will be discovered and uncovered, for it is part of the journey. You cannot hide behind your shadows, forever. Remember, in the Christ story template, the devil, the adverse ego, calms down (goes away), when rightly seen.

Also, the more you spiritually develop, the greater the energy of your temptations become, and the more subtle are the energy of the temptations. The measure of your higher dimensional energy dynamics application is the measure of your power to demonstrate and overcome temptation. *"All things work together for Good..."* (Rom. 8:28)

The experience in the wilderness initiates the ability to call forth your faculties, your disciples in greater proportion, for support in your efforts to demonstrate. The little self gives your experiences great meaning, and uses them to support you in excusing why you show-up the way you do. *I am this way because of my mother, if you knew her you would know why I am like I am!* To overcome this will take support.

The root word of *personality* is *person* and it comes from the Latin, persona, or mask. A mask is used to hide something. *"What masks do you wear? What masks and shadows still lurk beneath the surface in your subconscious mind, and still run your life?"*

Masks that the adverse ego, or the little self, wear come in many forms. The interesting thing is that when you have a mask or persona on, you think no one can see through it, you think you are hiding well! Wrong! Most of the time others can see your mask or persona, clearly.

Masks and personas show-up in forms such as a desire to be in control, playing the victim to get attention, being sick, being strong, being rich, wanting attention, labels you take one like a banner (mother, father, President of the Board, CEO, etc.) and much more.

During my wilderness experience (one of many), which was a devastating experience, I had a dream in which a dark-skinned came into my house. She invited me to take her hand, and lead me into the basement of my house, which I did not know existed. There was so much activity going on. Many, many other dark-skinned women were working down there. She said, they were making, an elixir of miracle grow. She asked if was ready to partake. I said, yes, and the journey with feminine nature of the divine began. Not long after this, I was gifted with unpublished writings of Mr. Fillmore, on his journey with feminine nature of the divine; and I was also introduced to the Black Madonna. Each one in its own way, turned my life upside down!

You do not know your true dominion and majesty in the Kingdom of Heavens until you successfully overcome a trial or a temptation, until you demonstrate your spiritual skill-set, and set your "demons" straight. Your power lies dormant until demonstrated, until that which is non-reality is transformed into authentic spiritual principle.

In the Christian Testament scriptures, there are three major temptations that occur during the wilderness experience. Per, Mr. Fillmore, these three specific temptations are ones that every person who sets forth on the spiritual journey will meet because; *"they [the temptations] represent the desires and ambitions of the untried and untrained forces still in the subconscious mind...They are allegorical representations of the way in which certain states of mind are handled by the initiate."*

The three temptations hold a desire for something outside yourself, and thus have the capacity to take you off course. The three temptations are:

1. Power   2. Wealth   3. Pride

1. _Power_ as a temptation holds two spectrums of energy, both are forms of wanting to be seen, and are also, an abuse of power. First, wanting to be seen as powerful, key word seen. This temptation reveals itself as: a desire to get, take or have something you think you don't have, at any cost; to be seen as having power over others; to be seen as the one in control. This is the one that reveled to Jesus, by the devil, his adverse ego. I will give you all this, if you just...
    On the other end of the spectrum, the second form of power as a temptation is unconsciously being seen as a victim of life. To be seen as having circumstances that are out of your control, in order to receive pity, or recognition through negative projection—feel sorry for me, poor me—etc. These are energetic forms of a need for power and require transformation.

2. _Wealth_ as a temptation holds the energy of a desire to be seen as being rich or of high status. You believe life is about money, rank, prestige, and status, and you want it. Remember, wealth is not about being rich, for there is nothing wrong with having wealth; it becomes a negative, non-supportive energy, when you are willing to sacrifice _who you are_ and all you stand for, in order to get it.

3. _Pride_ as a temptation holds the energy of self-righteousness; always wanting to be seen as right, entitled to favors or arrangements, or proving how wonderful you are in order that you might be recognized by others as being special or having special gifts.

These are the types of energetic temptations or "error-adverse ego-survival-oriented" states of mind that must be overcome. The "demons," or errors thoughts, perceptions, and beliefs in mind must be transformed, cleared, or released. This only happens through the demonstration of spiritual principle.

The final scene of wilderness experience, is after the temptations are presented, and Jesus says, _"Get the behind me Satan."_ You no longer have power here! When you are complete in the wilderness, the energetic illusions held in place by duality thinking disappear, for they have no substance to sustain them. All truths and principles revealed or taught by Spirit must be demonstrated, as a part of practical application.

You have concentrated on growing into a greater dominion and awareness of the I AM, you have been tested by your own adverse ego, and you have proven your dedication to Spirit through overcoming and demonstration. In victory of your demonstrations over temptation you return to the world armed with the power of Spirit. Know that at every level, all forms of temptations must be met, overcome, and transcended before you can move onto the next initiation.

The Energy Supporting the Ultimate Learning from the Temptation Experience—There is Only One Presence and One Power Active in My Life!

The energy held in the silence of the desert and fasting from things of the world strengthens and reveals your power and ability to overcome the dense energy of temptation. Temptation is created by your own adversarial thoughts (the devil). After new thought energy and power is gathered and proven, the angels (new and higher thought-energy), come to minister to you. These thoughts grow and ground you, even more.

Joel Goldsmith said that by age 30, or as long as it takes to perfect it, perhaps even lifetimes–through the power of Spirit working in and through you—there evolves a new level of awareness and skills. You now automatically *recognize the nature of error,* the nature of human-kind, and the non-reality of the "illusions" of the world.

In this way you grow out of the consciousness of good and evil into knowing One Presence—One Power —One Energy, as the only energy and power inherent in all things.

In duality consciousness, you see the energies of good and evil, good and bad, right and wrong; a sense of separation and duality is "alive" in your consciousness. Through continued study of spiritual principles and practical application of those principles, your thoughts are renewed and regenerated.

You no longer try to or desire to solve the problems of the world, at the same level that created them. You realize your truth; you are now conscious and thus, aware of what you are doing with and through the mind. You are now using your thoughts and actions in service to Spirit-Law-Principle. Through the Wilderness experience you overcome duality and the polarities of good and evil.

Overcoming builds character. It doesn't mean that bad things don't happen to you; it means that you see them differently and respond to the differently. You have now developed the capacity not to struggle or resist when confronted with difficult situations, people, or negativity. Remember, negativity or "evil" disappears in the face of non-resistance, for evil is not a power.

When you put the "devil" of little self behind you, you are free of the mind-conditioning that started everything in the first place. This is the spiritual reality behind the story of the wilderness and the temptations in the process of regenerating the mind.

Regeneration, bringing all of the forces of the mental, emotional, physical, and human elements of the soul under the auspices of the Christ. When the self, little i, "devil" ceases to operate, higher dimensional realities which include love, wisdom, creativity, and more come alive in your being. The *"angels come to minister"*. Thus, when the self finally gets out of its own way; *"the way is prepared,"* for the next level of evolution. The higher thoughts come to prepare and support you in the next phase of your unfoldment and ultimate demonstration through sacred service, a new mission, a new ministry of demonstration, and transfiguration.

However, there is still more to learn, express, overcome, and demonstrate before a full transfiguration in consciousness occurs. Know that from this point forward, if you have any remaining shadows or secret life, they will be exposed by "light" of higher awareness.

Temptation, and its variety for forms or tests, are often subtle, and take forms you may not instantly recognize, like that of illness. Yes, illness! You must stay alert and aligned with higher dimensional awareness in order to observe the energy in front of you and to make conscious choices, from a higher level of awareness so that you do not slide back into old ways.

From this point on, you will begin to notice that things of the material world have lost their power to seduce you or to produce a lasting negative reaction. Why? Because there is no longer any energetic substance to sustain their energy, thus, they must disappear. You have demonstrated your proficiency of spiritual principle, and now you are ready for the next phase of evolution and its experiences. You are ready to demonstrate, what you have established in consciousness. You are ready to do service work, in the name of demonstrating "Christ Consciousness."

4ᵗʰ Initiation: Demonstration —>Transfiguration

The 4ᵗʰ initiation is Demonstration, demonstrating what you have learned through ministry; sharing a sacred service with the world. It is a time where you practically apply, through action, what you have established in consciousness. You have now dissolved many of your 3ʳᵈ dimensional fear-based ideas and ideologies and are ready to share with others what you have gleaned and absorbed.

This initiation culminates with a full transfiguration of the mind and body, and the Higher Self-I AM assuming dominion in consciousness. The energy of old ways of thinking, feeling, and being is no longer in charge, the majority of your energetic and vibrational forces have been transformed into higher levels of awareness.

The Energy Supporting the Calling of Your Disciples

Before you can really begin your demonstration of spiritual principle through sacred service—teaching, preaching, healing, etc.—you will need to call forth the energy of your various faculties, centers, or disciples. Your disciples are represented by energetic centers within your body system that must now be trained – you are entering into alignment training. Disciple means, one in training, one who is consciously developing spiritual mind discipline and aligning their energies through strength of character.

Before the wilderness initiation, your twelve faculties or centers functioned at the metaphysical level of the letter of the law—*an eye for an eye.* You have now shifted from a consciousness of an *"eye for an eye"* into a transformed awareness of One Presence and One Power. You have initiated an understanding of no one, and nothing is against me or can hurt me. You now hold a state of mind that is no longer energetically or vibrationally in alignment with fear. Awareness and alignment are developed through a deeper understanding of what non-judgement, non-resistance, and non-attachment, truly mean.

Your primary guiding principle now, is a state of awareness, which I call, conscious neutrality. It is a state of detached compassion, and a sense of conscious neutrality. This does not mean that you do not care—it means that you no longer carry any attachment to old conditioned energies to things and beliefs.

Because you are living from a more disciplined mind awareness, conscious neutrality, you are releasing the need to judge outer events, situations, and even people based on good-bad, right-wrong. You are moving towards a continual stream of spiritual discernment and guidance.

Spiritual discernment invites you to ask different questions about what is happening around you. Questions such as: "Does what is going on matter, or apply, to my spiritual journey?" "Is what is going on, in the highest interest of the spiritual journey?" "What is greatest opportunity for a lesson here? These types of questions put all outer events into a new perspective. You are now aware that it is definitely, not about how much you know, but about how much have you absorbed, appropriated, assimilated, integrated, embodied, and established in consciousness.

The Energy Supporting the First Disciples Called

The first group of disciples called forth, as Jesus starts his ministry of sacred service; are the fishermen (Mt.4); Peter (Faith), Andrew (Strength), James (Judgment), and John (Love). The other disciples are called shortly after them.

"When people begin to follow Jesus in the regeneration, they find that they must cooperate with the work of their disciples or faculties. Heretofore, they have been under the natural law; they have been fishers in the material world...call [your] faculties out of their materiality into their spirituality [as Jesus did]."[37]

The disciples, through a conscious training of spiritual mind discipline, are transformed and lifted into new levels of higher dimensional awareness. Thus, they will shift from being fishermen in the material world, to being fishers of men. *"Fishers of men: spiritually quickened...strongly fortified in Truth and able to help others to find the light."* [38]

The Energy Supporting Faith

Peter-Faith as an energy is a grounded trust founded in the working of Law, at a higher level. Faith is the perceiving power of mind, linked with the power to shape substance (imagination). It holds the capacity to conquer doubt, through a firm knowing of spiritual Principle. In todays more quantum thinking, Peter would support you in shifting from the Law of Cause and Effect, to the Law of Causing Effect.

---

[37] Charles Fillmore, *Twelve Power of Man*, Unity Publishing
[38] Charles Fillmore, *Revealing Word*, Fishers of Men, Unity Publishing

The Energy Supporting Strength

Andrew-Strength as an energy is an unwavering stability of character, and a knowing confidence and depth of understanding of spiritual principle you are expressing. Strength evolves the courage to do what is right and needed in the moment, to overcome temptation.

The Energy Supporting Judgment

James-Judgment as an energy is spiritual discernment, the listener and hearer of intuition, and the initiator and caller of the fires of Holy Spirit. James holds the capacity to stand firm and make right choices through wisdom and spiritual discernment.

The Energy Supporting Love

John-Love as an energy is rooted and grounded through wisdom and is the strongest force in the Universe. It is a state of transforming power grounded in harmonic resonance and balance. Centered in unconditioned Love, you are a transforming power. There are three called before love, this means something. You cannot just jump to love, you must do other conscious work first, before love is called, otherwise it is not a grounded understanding of love. It more of an airy-ferry, its all love!

These four are called first for together they have the power to create a systematic, organized, and progressive movement of energy that works with intention, vibration, and law to attract to you, the highest level of possibility in accord with energy and vibrations.

The Energy Supporting Healings and Miracles

"The Spirit of the Lord is upon me and I am called to heal the sick, restore sight to the blind, cast out demons, and reform the sinner..."

In Matthew 4, Jesus shares this quote from Isaiah. He states that it is his mission, his sacred service, in support of humankind. And he now goes forth to demonstrate what he has established and appropriated in consciousness, to date.

Remember, every detail of the life of this master teacher, is a symbolic point that reveals what you too must grow through and into, albeit, your experiences may look different. Every detail of his journey is a part of the Christ template for soul evolution. Every detail and element is a metaphysical revelation that shows the way towards ascension into Christ consciousness. And remember, Jesus shares that everything he accomplishes you can do, also.

Do not be fooled by the words, ascension, higher, elevated, etc., for they are just descriptors. All levels of energy and dimensions are present within you now; all levels, all dimensions, are here right here, right now, in you; it can be no other way.

Also, have you noticed that healings are occurring in Jesus' presence and he has not, yet, taken on the fullness of Christ consciousness? What does this mean for you?

When Jesus said, these things I do you can also, he meant it! You do not have to be in fullness of the Christ consciousness for healings and miracles to occur in your presence; you only have to be willing to claim the fullness of the Christ consciousness, as who you are with an unwavering faith. When you do this wonder-filled healings, miracles, and synchro-divine events will take place in your presence.

The Energy Supporting the Sermon on Mount: Introduction, BE-Attitudes, and Teachings.   Matthew 5 – 7:19

The *Sermon on the Mount* is a mystical teaching that holds a timeless blueprint and instructions on how to be a master facilitator of energy—remember, its all energy—through the development of; spiritual mind discipline, vibrational alignment, non-resistance, prayer, forgiveness, being a non-anxious presence, demonstration of spiritual principle in consciousness, and a fourth- dimensional state of awareness referred to as, conscious neutrality.

When seen through the lens of higher dimensional frequencies and awareness, the *Sermon* reveals mystical teachings that support you in energetic and alchemical transformation. Alchemical transformation is the changing of energetic substance from one form to another. The patterns of form shift (moisture → water → ice) and the substance appears different or changed, however, the underlying substance, which is energy, can never be destroyed. Thus, an alchemical transformation in consciousness is a movement of energy that creates a vibrational shift into a different level of awareness.

That shift in consciousness is out of third-dimensional limitation anchored by the adverse, survival-oriented ego into a greater knowing and realization of who you are (BE) as Christ-illumined-enlightened-fifth dimensional (and higher) consciousness—a new state of harmonic resonance and balance—conscious neutrality—the presence of "all is well."

In my own research on the *Sermon on the Mount*, the *Sermon* revealed to me a hidden, mystical teaching. It unveiled a portion of the blueprint for the expansion of consciousness into higher levels or dimensions of awareness; along with a step-by-step how-to-instructional curriculum on how to live, demonstrate, and be a master facilitator of energy, vibration, and frequency.

This is my attempt to share the results of the alchemical transformation that took place in me as I unwound the mystery, wisdom, and instruction of the *Sermon on the Mount*. I discovered, from my perspective, that the intention of the *Sermon* is to initiate a focused, conscious, orderly, systematic, intentional, progressive, and evolving movement of new ideas and levels of awareness in the mind regarding soul evolution, and the subsequent mastery of energy facilitation.

By saying yes to engaging with the metaphysics, energy, and mysticism imbued in the *Sermon,* and the whole *Gospel* experience, you are embarking upon a journey, a journey of intentional transformation, transmutation, assimilation, absorption, shift, and change. It will result in a change in mind, body, DNA, consciousness, experiences, thinking, feeling, and more. Your willingness to participate fully and resolutely has the capacity to alter your life, both internally and externally.

In the quadrilogy of soul mastery, the development of the 4th dimensional state of conscious neutrality is what hones the qualities and elements required for being a master facilitator of energy, a master of abundance, and a master of the law of attraction. The elements of the *Sermon,* metaphysically interpreted, reveals how to move into that 4th dimensional, and higher, awareness!

The metaphysical concepts found in the *Sermon,* clearly reveal how spiritual principles—life, love, wisdom, peace, abundance—can be energetically mastered, demonstrated, and manifested in all areas of your life. Once established and demonstrated in consciousness, they energetically permeate all your actions, words, and thoughts and you become a conscious energetic force for good. However, for this to occur, the ideas and principles requires engagement in a conscious and orderly way, which will result in an expansion of awareness. As you engage in the metaphysical energy underlying the *Sermon on the Mount,* view it from the context of the possibility that the *Sermon* is a two-part teaching.

The following paragraphs are an overview of the two parts, with the full teaching following. The two parts are;_Part 1, MT. 1: 1 -20 and Part 2, MT. 5: 21 - MT. 7: 19.

Part 1 is the Beatitudes. The BE-Attitudes are actually the *end result* of who you will be, once you do the work described in Part 2 (the how-to-do-it message, which starts at MT. 5:18). The individual and specific attitudes, qualities, and values represented in the BE-Attitudes, are the end result, the revelation of the new state of dimensional awareness or consciousness that will be established in mind, body, and being through the raising of your energy, vibration, and frequency by having mastered Part 2 of the *Sermon*.

The name I was given for this new state of awareness—is conscious neutrality. Once conscious neutrality is established as a state of being, who you BE, you will notice that you are engaging life as a master facilitator of energy (refer to the Conscious Neutrality chapter for more information). However, this state of being only occurs after you first do the spiritual and emotional maturing work as outlined in Part 2.

Part 2 shares a step-by-step, how-to plan for releasing energetic density and for transforming the third-dimensional, fear-oriented, adverse ego. It is a blueprint on how-to spiritually and emotionally mature, in order to alchemically transform consciousness and BE a master facilitator of energy.

Part 1 - Introduction and BE-Attitudes – MT. 1: 1-20

Introduction
MT. 1: 1-2

1 Now when Jesus saw the crowds, he went up on a mountainside and sat down. His disciples came to him 2 and he began to teach them.

1 Now when Jesus

Jesus, a man, a person just like you and me, a master facilitator of energy, and teacher who innately understood the mystery of and demonstrated the mastery of how to consciously engage the higher dimensional aspects of energy, vibration, and frequency.

saw the crowds A gathering of negative and lower energetic thoughts and feelings within his being, desiring to arise. Those thoughts that would say – who are you to do this?

he went up on a mountainside

He consciously made a choice to lift and shift his thoughts into alignment with a higher state of energy, vibration, and frequency. Into fifth dimension and multidimensional awareness, where the energy and constructs of "One Presence and One Power" reside.

and sat down.

He surrendered his small-self thoughts, his adverse ego, and allowed an alignment with higher energy and frequencies to shift his thoughts.

His disciples

Energetic fields of thought/feeling within his cellular structure, that have already been prepared and raised into new levels of awareness.

came to him

These prepared centers of consciousness within that are already in alignment with higher awareness and frequencies, support an alchemical transformation of lower energies and a movement into alignment, attunement, and harmonic resonance with the "Christ" vibration. A vibration that is always present in your field of awareness.

2 and he began to teach them…

Now, being in a state of alignment, Jesus was receptive and began to receive new downloads of light/information, which his centers of energy/consciousness/awareness were receptive to receive. Aligned to higher vibrations and frequency, the Universal Impulse moved through him and created an expansion of consciousness within the entirety of his BE-ing.

<u>Beatitudes</u> – BE-Attitudes — BE!
<u>MT. 1: 2-17</u>

2 He said: Blessed are…

> **Bless**, metaphysically means to confer energy as the activity and action of God/Spirit in motion and to consciously increase and multiply its capacity for working for the higher good for all.
>
> > **Bless**, if "bless" means to confer energy upon, consider what type of energy you might be blessing your life with. Is it a blessing of conferring the energy of resistance, or higher dimensional activation? The law works: right use or reverse use, the law works!
>
> **Are**, a statement of completion, already complete, already established. Thus, are means, AS one already in this state of awareness.
>
> **Blessed are,** as you are this walking, talking energetic state of Bless, you are consciously conferring the energy of God activity in action toward increase and multiplication, in every interaction you engage in.

Blessed are those who are an increased speed of vibration, are cellularly immersed in an energetic vibrational field of higher dimensional awareness, luminosity, and light and they confer that energy upon of like resonance and/or are receptive to it.

Blessed are those who be a master of energy for they have the spiritual principles required to perform energy work at higher levels and dimensions of awareness. In this state of soul mastery, there is now a demonstrable, ever-present, repeatable, and sustainable energy available. And it is instantaneous when called upon in the moment. Once the establishment of the demonstration of a spiritual principle (e.g., abundance) in consciousness is established, it is sustained because new neural patterns have been rewired into the brain's neural pathways and they now serve as the brain's default mechanism.

3 Blessed are the poor in spirit, for theirs is the kingdom of heaven.

Blessed are

Now, vibrationally you are the energy of God activity in action moving all things toward increase and multiplication.

the poor in spirit

Those who have made themselves a mental vacuum, to be filled with Source. Those who are surrendered, who are willing to seek and do the will of Universal Presence/Spirit/Source/ Energy/ Invisible Intelligence.

For theirs is

Theirs IS now.

The kingdom of heaven

Higher dimensional awareness, the unified field of consciousness, the connector of all potential and possibility working through the activity of law.

4 Blessed are those who mourn, for they will be comforted.

Blessed are those who mourn

To mourn is to desire, to long for, and cry out to be with one you love and miss. You long and cry out for an expansion of higher dimensional understanding and awareness of who you are. Those who cry out and long with all their being for an expanded sense of knowing God/Spirit/Universal Presence are compelled by an energy or desire for universal awareness. Those who cry out saying, "I know that I don't know, but Spirit knows" will call forth resources to be utilized as part of the comfort.

for they will be comforted. They will be guided to sources and resources in order to be alchemically transformed and to gain higher dimensional awareness, energy, joy, etc. Which is the true treasure.

5 Blessed are the meek, for they will inherit the earth.

Blessed are the meek

The meek, the humbled, those who have energetically discovered and overcome the needs of the adverse or lower ego, and allow their thoughts to be receptive to spiritual realities, only.

Those who understand sacred service as a state of being—sharing their talents and time—not to be seen—but for the glory of good for all.

For they will inherit the earth.

They will embrace a new vibrational level of awareness within their field of energy.

6 Blessed are those who hunger and thirst for righteousness, for they will be filled.

Blessed are

Bless is one who is now the energetic vibrational field of a higher dimensional awareness— for they have the capacity to confer the energy of Spirit.

Those who hunger and thirst for righteousness

Hunger and thirst: those who have released duality thinking; who seek Source in all experiences; who ask, long for, and live from an unquenchable desire for expanded awareness of "right-thinking."

For they will be filled.

They will be higher awareness in action, conscious neutrality in action, and a master facilitator of energy and spiritual principle.

7 Blessed are the merciful, for they will be shown mercy.

Blessed are

One who now holds the energetic vibrational field of a higher dimensional awareness.

The merciful

Those who are willing to forgive, energetically releasing resistance and a negative force field.

For they will be shown mercy

They will send forth a new vibrational signal into the unified field of consciousness that will return to them, and those they "pray" for, greater good.

8 Blessed are the pure in heart, for they will see God.

Blessed are

One who is now the energetic vibrational field of a higher dimensional awareness.

The pure in heart

Those who dwell in the presence of Source, for they have put away thoughts of a "lower" nature, lower vibrational energy fields anchored in the third dimension: fear, anxiety, me first, playing small, greed, jealousy, insecurity, etc. For they are now receptive to purer, clearer, higher vibrations and frequencies and hold pure and intentional energy for highest good of all to be shared and expressed.

For they will see God.

They will be a higher vibrational awareness in action. You are the place where God is expressed in the world, you are God in expression; those who see you see the energy of God expressing or not!

9 Blessed are the peacemakers, for they will be called children of God.

Blessed are

One who is now the energetic vibrational field of a higher dimensional awareness.

The Peacemakers

One who has raised consciousness to a state of conscious neutrality and the presence of "all is well," regardless of what is happening in the energy field around them! And in that presence of higher dimensional awareness, energy shifts and transforms.

For they will BE called children of God.

They will be recognized and acknowledged in the higher realms as one in alignment with I AM, the higher awareness of being, not necessarily in the realm of the third dimension. For in the third dimension, their gifts and talents may still be misperceived and not recognized.

10 Blessed are those who are persecuted because of righteousness, for theirs is the kingdom of heaven.

Blessed are

One who now holds the energetic vibrational field of a higher truth.

Those who are persecuted because of righteousness

Those who have matured the faculty of Strength; and thus, have the strength to be persecuted, rejected, energetically-physically-emotionally misunderstood, and not seen for positive, higher awareness field of energy that they carry.

Righteousness is "right" energy—having aligned with a higher awareness. This may even be revealed an internal struggle with small self – who do you think you are? You are not worthy to do this!

For theirs is

It is established. They now resonate harmonically with the greatest riches, the gift of higher frequency awareness.

The kingdom of heaven.

Higher dimensional awareness.

11 Blessed are you when people insult you, persecute you, and falsely say all kinds of evil against you because of me.

Blessed are you

One who now holds the energetic vibrational field of a higher dimensional awareness and is able to confer that energy to those who are receptive.

When people insult you, persecute you, and falsely say all kinds of evil against you because of me

Often, people – including your own internal critics and "thought people"– do not resonant with the energy field that you now hold. They do not harmonically resonate with the energy, they fear and misunderstand it, and thus, reject it – and you.

12 Rejoice and be glad, because great is your reward in heaven, for in the same way they persecuted the prophets who were before you.

Rejoice and be glad, because great is your reward in heaven

Regardless of appearance, of how people treat you, stay focused and intentional. Be, and stay, energetically and harmonically aligned and in resonance with higher dimensional, for that is where true joy and satisfaction reside.

The reward of choosing to stay in the state of conscious neutrality expands the vibration and frequency of the consciousness of truth that is anchored and stabilized within you. The more you consciously work with it, the more embedded it gets.

For in the same way they persecuted the prophets who were before you.

Other master facilitators of energy who have come before you have had these experiences and have persevered in the higher realms.

Salt and Light
MT. 5: 13-16

13 You are the salt of the earth.

Salt means energy of higher dimensional awareness and "truth": having taken on this state of awareness, conscious neutrality, you are now the energy of higher dimensional awareness in expression. Now you must be vigilant to stay in that consciousness, for without higher dimensional awareness as the expression in consciousness, nothing you say or do impacts your energetic field or the fields you play in with others.

14 You are the light of the world.

Light means higher frequency, awareness, holder of information, and energy in motion. When you are aligned in a higher level of light frequency, It cannot be hidden for It is a radiant, luminous energy.

That luminous energy will express Itself in a multitude of ways, such as transforming your current landscape of reality and the energy in every cell of your being as It radiates out through your energy field.

People will sense It: some will be attracted to it and some will reject it out of fear without understanding why they are rejecting It.

Higher Righteousness and Fulfillment of the Law
MT. 5: 17

17 I have not come to abolish the Law ... but to fulfill it.

To **fulfill the law** means to raise the level of energy, the field of vibration in which you live, move, and have your being, in order that you may BE a state of awareness that effects change by virtue of your presence. Righteousness means "energy that has been made right, lifted to new states of vibrational awareness." You are in "right energy" when you shift your awareness from third-dimensional constructs to higher levels of awareness. You currently reside in third-dimensional constructs that are steeped in fear, anger, greed, conditioned love, etc., and the work is to raise your vibrations in order that you may be a vibrational force field of higher dimensional awareness.

This ends the segment known as the Beatitudes. The BE-Attitudes are the end result who you be once you do the work to release the hold that the third-dimensional adverse-ego, with its false beliefs and perceptions, has on you. The adverse ego, a third-dimensional construct, holds you hostage to its wants, its justifications, its need to be right and to be seen as right, but as you begin to grow into higher dimensions of vibrational awareness, that egoic hold begins to dissolve.

As you ascend in awareness, the veils of egoic illusion dissolve, and who you are begins to alchemically and energetically transform. You shift energies and vibrationally start to be the tenets of the Beatitudes. You be a new state of awareness in regard to being "poor in spirit," "meek," "a peacemaker," and more. You alchemically transform the lower states of limited, adverse-egoic awareness into new dimensions of understanding.

As conscious neutrality in action, you are now a receptive energy field of multidimensional awareness, a state of being that higher frequencies move through effortlessly, as long as you remain in alignment. In this higher state of vibrational being and awareness, a fourth- and fifth-dimension orientation, you are a blessing. And the energy of higher dimensional awareness goes forth, through you, to do its work of transforming energy for the highest good of all involved.

In this state of higher energy, you *are* the presence of all is well wherever you go. People feel it and that energy has the capacity to shift, lift, influence, and transform any negative energy present in the field around you. You become an influencer of the energetic field, not just yours but others.

Part 2: How-To-Do-It. How to Develop Conscious Neutrality as a State of BE-ing.   **MT. 5:21—>MT. 7:19**

Anger, Murder, Reconcile Quickly, and Release on the Altar MT. 5: 21-26

Murder, anger, and contempt are third-dimensional energetic fear-survival oriented constructs that eventually must be transformed, those energies must be aligned with higher dimensional awareness. This takes spiritual mind discipline and the capacity to release ego needs to put God first.

To surrender to the highest possibility for the greater good of all, you must be willing to do the alignment work, even if you don't want to! You must reconcile/align old energies, lift and transform them. Twice, the master teacher refers to the idea of going into the "secret chamber of the heart" to pray. To pray is to consciously align with the higher energies of Source. And yet before you do this, you need to do the following:

Leave your gifts. Leave your good points, idiosyncrasies, your anger, agitation, and all requests at the altar and seek first the reconciliation, the alignment of all your energies. Before you ask for your good, you need to be in a surrendered place, willing to align to higher energy and to invite a new energy that is focused, intentional, nonresistant, unattached, nonjudgmental.

Forgive. Give for: align with a new harmonic resonance, align with a new energy, and give that energy forth instead in the face of perceived wrongdoing. And you can't do it just a little or just what you want to do. All must be aligned (the last penny given, no shred of energetic animosity toward anyone or anything).

Settle matters quickly. Release your 3rd dimensional vibrations of anger and upset and do it quickly, for the longer you hold onto the energetic strands of lower vibrational thoughts and feelings, the more potential they have to create habitual ruts in your neural pathways—ruts that become embedded beliefs and perceptions that unconsciously rule your life and hold you hostage.

And there is no stronger prison than your mind when it is stuck in the muck and mire of adverse ego perception, a perception that believes that someone or something is against you!

Freedom starts from within; it does not happen because of conditions occurring in the outer world. Freedom is an inner-world process.

Last penny paid. You cannot energetically pretend to have released the lower vibrational hold of negative thoughts and feelings. You must let go and be a clean energetic slate.

Have no resistance be a place of focused intentionality for the greater and highest good of all, not just self.

Adultery and Right-Eye, Right-Hand Release
MT 5: 27-30

To **commit adultery** energetically means to make a conscious choice in the moment to move away from balance and alignment with Source Energy and awareness to appease ego needs of being right, seen as being right, being first, being angry, not being transparent or authentic, playing small so others feel sorry for you, and so on. It also, metaphysically means, to water down the truth of things, in order to do things "my" way.

There is an energetic subtlety here that is important to mention about adultery. It is a deeper and more subtle idea in regard to the moving away from the alignment and balance with Universal Source Energy.

It is a subtle reference to your interior masculine and feminine energies and a reminder that they must always be in balance (and conscious interplay) with each other. Neither one can be able to dominate, regardless of your gender.

Remember these two things:

First, Jesus was teaching in mystery, only to be understood by those who had *ears to hear, and eyes to see*, those who were in higher dimensional awareness. Which was not the level of scribes and Pharisees, who believed they already knew!

Second, there was neither the language nor the understanding to speak of these things in the same language or capacity that we have today.

Adultery occurs when the **right eye/right hand** misuses or abuses its birthright of truth (right refers to masculine energies and left refers to feminine energies) or when masculine aspects of being, shun or move out of balance with their feminine counterparts and qualities of compassion, integrity, transparency, authenticity, connection, "both/and," etc.

We speak of *one* or *oneness*, but for sacred union to occur, the masculine and feminine energetic qualities of being must be in harmonic resonance and balance. You cannot cheat, one side or quality cannot be loved more than God.

Charles Fillmore said in 1916, "Jesus Christ reached the limit of his masculine consciousness. He was the heat of world, the light of the world, but he could not be this heat and light ... he had to take on his

Divine Mother consciousness. We have not yet the understanding of this Divine Mother consciousness."

And we too, must take on the divine mother consciousness!

This is an imperative in the evolutionary process and in the cycle of conscious evolution.

Do not dismiss this or spiritually bypass it.

Right eye, **right hand**, and **choose quickly**. As you practice self-observation, you will begin to notice when you are vibrationally out of alignment, when you are not focused on higher awareness. When your thoughts are straying (right eye or right hand), it is imperative for you to observe that you are out alignment and then quickly make a conscious choice to realign. For one thought leads to another and, before long, if you do not take care of negative thinking and feeling quickly, you will be steeped in negative patterning.

<u>Divorce and Marrying a Divorced Woman</u>
<u>MT 5: 31-32</u>

Divorce energetically pertains to the feminine and masculine natures of the divine. Your work as part of the process of evolution is to experience the sacred union of the masculine and feminine aspects of being (refer to the wedding at Cana for more insights on sacred union).

However, at times the feminine energy of compassion, authenticity, and transparency can get in the way of your ego's agenda, so you divorce from it and do what you want. Your feminine energies are scorned and set aside.

The feminine is often referred to in the following manner: *Who needs emotion? They just make you seem weak! Don't cry, it is not seemly.* However, when you are overly sensitive, get hurt easily, go overboard with your sense of compassion, act like you are victim of someone else's projections of thoughts, etc. you have divorced the strength of your feminine nature. The feminine nature is soft, and it is also fierce, when it needs to be. But one must be in balance, in order to know how to effectively wield these energies.

This last part is subtle: **anyone who marries a divorced woman commits adultery**. If you join in union with lower emotions -- lower survival emotions, which represent the unhealed aspects of the feminine nature and also emotion divorced from truth of being – and allow them to run your life, and you consciously do not work with them to try to reconcile or raise them—or you deny your emotions have any place in your work or life—then you are divorcing the feminine. All of which takes you out of balance. Not good: they will become part of your shadow!

These negative, out of balance energies will become embedded in your cellular system and support you in staying in a sense of third-dimensional separation—and **adultery** is separation from Source, being out of alignment with Source Energy, or not in harmonic resonance or balance.

Oaths and Retaliation
MT. 5: 33-37

Oaths. Energy is always shifting and changing, therefore do not say "ever" or "never," for you don't know what will be asked of or required of you. Stay focused and intentional on raising your energy and making choices based on what is making itself known in the moment. Swear to nothing, for you know nothing. Really, you don't. You only have perceptions in the moment, and everything changes so quickly, so how can we be sure of anything? Dance at the edge of mystery through knowing, "I know that I know nothing!"

Retaliate. Don't do it. It is a lower dimensional reaction and it always reaps bad karma. One living from the state of awareness of conscious neutrality knows what this means. Non-retaliation, or non-resistance, as an act of real and true power, can only occur when you have command over your emotions and responses.

Let your yes be yes. Two thousand years later and this is an incredible expression of how to imprint upon the unified field of consciousness and the energetic field of substance. When we imprint upon the field, as we do with clarity and intentionality, we are working with the law to bring forth the highest possibility that we are vibrationally in alignment with.

Eye for Eye, Go the Extra Mile, and Give
MT. 5: 38 - 42

Eye for an Eye. BE nonresistance in action, **give** love for ____, and go the extra mile in times of perceived persecution and peace; no more needs be said.

Check to see if your ego didn't like that.

Did it throw up any resistance such as, *"But what if? You don't understand this situation!"*

There is your clue! Resistance breeds more resistance.

You cannot expect to sow resistance and grow love!

Love for Enemies, Pray for Those Who Persecute You, and BE Perfection in Action
MT. 5: 43 - 48

Love your enemies. Energetically, this means seeing all persons as people who are working to open their hearts to find their soul. They are doing the best they can with the energy level they are currently working from. This doesn't mean you never take action but be assured that it is inspired action not an action that holds pity, anger, or a need to save them from themselves. They are where they need to be in order to learn what their soul is inviting them into.

Piety and Giving to the Needy, Do All in Secret, Know God Will See and Reward You
MT. 6: 1-4

Do not practice righteousness in front of others. The teacher says this several times in a variety of ways. So, take heed of the message. Humility! Practice it, practice it, and practice it some more, until it becomes a natural state of BE-ing.

Give in secret. Do not brag about what you have done or given to so and so. Be benevolent, be giving, be without words.

Wordy prayers mean nothing; the real question is, "Are you aligning with the luminous light, the dynamic energy within?"

Wordy prayers are their own **reward**, and **the reward of secret prayer** is that substance and light sustain it. You need the communion and union with the higher realms of being to tap this light.

Your greatest gifts come from silent sacred service, not from being seen for your service. For it is not you or I who does the work, but the energy of the Universal Impulse, the Universal Intelligence, or the God moving through you.

When you do your deeper work, establishing a new foundational construct for the ascension of the needy aspects of you, there is no need to shout it from the rooftops.

Just do your **work in secret,** for there will be **your reward** and your reward will be that you will rise in conscious awareness, vibration, and frequency and you will be transformed.

Prayer and Fasting, Go into Your Closet and Pray in Secret, and Father/Source Knows What You Need Before You Ask MT. 6: 5 - 8

As you go into the **inner room/inner closet** and **close the door**/leave all else behind—meaning as you enter the inner realms or inner dimensions — enter, bringing only the dominant vibration of your spiritual identity/I Am. Leave all other thoughts/feelings/emotions behind. This initiates the sacred secret journey to the interior of self – and it is dome through prayer and meditation. Start by going within without any agendas. Leave them all at the door, on the altar. Pray for nothing, no-thing, once you go inside. Issues don't exist in the fifth dimension anyway.

Meditation culminates in touching the stillness, even for just the briefest of moments. It is a dynamic state of mind or consciousness, that appears passive, that supports tuning in and aligning with Source Energy. It is an alignment or focus of intentional energy toward higher dimensional awareness.

Alignment with Spirit supports you being in harmonic resonance with the Source of your being, higher universal energies, which are always available to you. And as you do this, the energies will teach you, more and more, how to be in every moment.

All methods that bring you into contact with God are appropriate. Pray until you set up union with the One Mind, at which time the universal impulse will course through you and ideas will flow, concepts will be understood, energy fields become like books you can read, synchro-divine events will begin to occur spontaneously, around you, and you will synchronize with the energy of the inner quality of everything.

The Father/Source Energy knows what you need, before you ask, so concentrate on alignment, and the Universal Presence will create synchro-divine events to meet all your needs.

Per Charles Fillmore, as you pray and as you bless, you tap into and touch the inner forces of the natural world, especially the element energies and forces of the inner worlds (and through this, you have the capacity to alchemically transform things). With the loaves and fishes, Jesus tapped into the substance of their inner forces, for the inner forces of the electrons of two fish and five loaves were Spirit.

If you have this understanding of the mind, of the dynamic power blessing and prayer and of the mind to open the atomic structure of everything, you begin to see how Elijah started the never-ending oil supply for the widow. How Jesus took the young boy's fishes and loaves and expanded their substance. And how Jesus was able to rearrange the cellular structure of body and ascend into the heavens. You can do this too, for all these things were done under the law and through the law of blessing. They came through the understanding of the dynamic power of blessing and prayer.

In entering that inner closet, Jesus felt the dynamic power and energy of Spirit—*the action of God, the energy of God moving* — moving in him. He knew the law. *God is Spirit and they that worship him must do so in Spirit and truth.*

What is it that stops you from doing this? It is because of the adversary/the adverse ego/the small self. You have incorporated into your brain and body cells the energy of the adversary, the third-dimensional belief in fear, lack, limitation, not being enough, etc. When you start your journey into soul mastery, the adversary, those shadow aspects of self, must dissolve in order to allow more light into the cellular structure. The hell and agony that the third-dimensional shadow, and its perceptual constructs bring, must be left behind, dissolved into the nothingness that they are.

The process is like the chrysalis and the butterfly: To transform, the butterfly must kick and squirm out of the chrysalis and no one else can help. You must understand what is happening and do it yourself. Getting rid of old perceptions and the shadows that they have created is challenging work. It is not just platitudes and wishing it away or faking it until you make it.

Jesus went through a journey of spiritual evolution and of spiritual mind discipline, etc. just as you must do. He preceded you, and all of humankind, in soul evolution, evolving out of the old and into the new being in higher awareness: Christ consciousness.

The greatest factor in soul mastery is prayer. God is omnipotent, all power, but it is you who must release that power through and into your cells and being. Prayer is tapping into a higher principle and power to bring it into expression. In your prayers, in your concentration, in your meditation, silence, and speaking of affirmation, consciously desire to connect to the ONE source of being.

Pray in this Way, The Lord's Prayer
MT. 6: 9-13

Matthew 6:9 initiates the Lord's Prayer, with words similar to this; pray in this way. However, in the original Aramaic, the word underpinning that sentence is Beshemi. According to Aramaic scholar, Neil Douglas Klotz, the root word of Beshemi is Shem. Shem means, light, sound, vibration, or frequency. So, in light of today's Aramaic scholarship and interpretation, that sentence might now read; BE in alignment with the Light, and Vibration of the Energy of this prayer.

Now that you are energetically aligned, the work is to learn how to consciously and continually release resistance and to engage effortlessly and consistently with the higher vibrational energies of who you truly BE. Developing the capacity to allow might be the most important instructional demonstration you can learn. Why? Because you can meditate and align every day, but if you get up and go out into the world and allow discordant energy to create negative reaction within you, you have just negated all of the energy work you did in meditation.

When you allow Universal Presence/God/Spirit to be the leading vibration in your personal energy field, the law of dominion (dominant vibration)/the law of vibrational attraction then goes forth into the unified field to match with the highest possibility that is in alignment with your dominant vibration. And that possibility is then translated into a tangible manifestation in your life!

BE a harmonic resonant field of pure energy in expression.
BE the space of non-resistance, non-attachment, non-judgement.

Forgive and Fasting.   MT. 6: 14 -18

Forgiveness has nothing to do with an outer act. This has to do with a state of inner being. Regardless of what occurs, BE a state of non-resistance, non-attachment, non-judgment, conscious neutrality, and that energy will work in accord with the law to support you in all your endeavors. Right use or reverse use, the law works!

This is not talking about abusive situations and enduring physical abuse from another. No, this is about having the inner strength to walk away from the situation without blaming the other person. I am leaving because it is best for me, not because of another's faults or behaviors. Can you see the subtle difference there?

Fasting when you are engaging in energetic healing practices: don't brag about how good you are or what good fortune you are receiving because you are doing it so right. **Fast**, use little amounts of words. Allow your presence to be what is seen. And be present to the energy in the field around you, regardless of what you are doing or whom you are engaging with. If you notice that energy is discordant, you can **in secret**—without a word—begin to consciously shift and alchemically transform the discordant energy by virtue of your presence. This means you do not make a big deal out of your spiritual practices or gifts. Yes, you have them—so use them.

By **fasting** from bragging or engaging in egoic behaviors, you become known by your fruits/good energy, and that is the real **reward**. So, you don't have to tell everyone how good you are; higher energy speaks for itself.

Anxiety, Treasures in Heaven, Eye Lamp of Body, and You Cannot Serve Two Masters
MT. 6: 19 - 24

Treasure: Where your thought-feeling energy (**treasure**) is focused (on lower or elevated energies) determines the energy you bring to a space and how you engage with others. For **where your heart** is reveals your state of emotions (survival-egoic or elevated). Energy doesn't lie!

If your energy is not in a higher level of harmonic resonance (**your treasure**), people will hear it in your words, see it in your actions, and they will know that something is not quite right. And if they are sensitive, they will know that incoherent energy is present. Even if you think you are doing a great job of faking it, know you are not!

Eye Single: being a master of energy facilitation requires that your **eye be single**, that you develop a single-focus consciousness, a laser-like intentionality and generosity.

The **eye** is how you see and perceive things in life. Do you see through **dark**, unhealthy (cynical, defensive, etc.) **eyes**? Or do you see through **healthy eyes** filled with **light**, which see all is for good, regardless of appearance?

You cannot serve two masters. The Universal Impulse/nonphysical, pure positive energy/Christ consciousness/I Am is always aligned with and advocating for your highest and best. This energy is a powerful force of attraction that is always calling in the cooperative components required for that which you are desiring or asking for. It is the energy of joy, "all is well," conscious neutrality, unconditioned expression, and satisfaction.

When you are in alignment and allowing, the feeling is one of expansion and all is well, versus always needing more or a needy energy. Needy energy, however, it manifests, creates vibrational discord. Whenever you feel needy or negativity—when you feel the need to be sure that you are safe, to be taken care of by someone else, anxious, fearful, doubt-filled, etc.—know that you are out of alignment with your inner BE-ing! When you push against anything, you create resistance, and that resistance lowers your energy and creates a new dominant vibration.

Your dominant vibration creates the point of attraction for the vibrational signal that you radiate. This signal is what goes forth to do its work in the unified field of consciousness and possibility.

You are always in relationship with your Source, but the type of relationship is up to you. You are the one who allows things to influence that relationship, to establish a fear or faith vibration. What type of vibration do you want to be the energetic influencer of your reality?

You cannot serve two masters; you must choose which one will be your dominant vibration!

Allowing a negative thought and feeling to be the dominant vibration of your being influences your vibration and deactivates the positive energy you had flowing. Yet you have free will: it is your choice! How can you quickly shift energy? Ask these quantum questions:

- Is this an influence (**master**) I want to perpetuate (**serve**)?
- Is this an experience I want to keep having?
- What active, dominant vibration do I want to experience now?
- What master vibration do I want to serve, Spirit or ___?

Do Not Worry, Take No Thought, and Seek First the Kingdom
MT. 6: 25 - 34

Seek first the kingdom. Be in alignment with a higher vibration, lift your eyes to a higher realm of awareness, a higher resonance in the multidimensional spectrum. Seek first the kingdom and consciously shift to something more interesting. Do that and THAT energy will respond in kind. It is the law!

Do not be anxious. Take no thought. Be not concerned with what is going on around you. Look at all things from an energetic perspective. This does not mean you do not think; it means that you ALLOW your thoughts to be oriented toward action to be inspired by Spirit/Universal Presence. When you are in energetic alignment with higher vibration, that vibration goes before you to prepare the way, make straight the path, and to rearrange the matrix (an environment in something grows). You do not have to take any thought; the higher dimensional frequency does the work through you—It performs an alchemical transformation experience that occurs right before you.

Do Not Judge, Not Seeing Log in Your Eye While Seeing Speck in Another's, and Do Not Throw Pearls Before Swine
MT. 7: 1 – 6

Judge not, lest you be judged. Consistently learn to shift your energy to a higher vibration and awareness when things appear to be going awry or you do not like what you see. Stop giving meaning to things and stop being a made-up meaning-maker. You do not know the reason for everything.

Do not see the annoying speck in another's eye (energy field) until you have requested to see how you carry this same energy. We are mirrors for each other.

Do not throw pearls before swine. When you are engaging in a field of perceived incoherent energy—holding an energy of higher awareness and you sense it, see it, know it humbly, and yet others do not see it that way—shake off the energy and move on. Don't take it personally, don't try to change them, don't force an issue. There is nothing personal about it. It is an energetic dynamic of two different types or levels of energy fields colliding. It is not right or wrong. It is about readiness and being ripe for change and new levels of information.

Do not waste energy trying to convert others to your way of knowing. If they are not ready, they cannot see it through the same lens that you do.

Don't fight, resist, retaliate, or try to be right. Instead, shift your energy to higher levels of awareness. And remember, every time you try to convert someone to your way of thinking or feeling, you actually lower your vibration to meet theirs. Ponder this!

Having aligned in this consciousness, your actions will be inspired by higher dimensional awareness and frequency and will attract alchemical transformation experiences, synchro-divine events, and the highest possibility in the unified field of consciousness.

Know that the higher the vibration you consciously send forth, the more impact it has on the field and the more you add to the overall consciousness of greater good for all, as a collective whole.

As you are alchemically transforming your life, you are imprinting a higher vibration onto the field and supporting the capacity for a new awareness in collective consciousness, regarding the establishment of spiritual principles in mind (abundance, etc.).

Ask and It Will Be Given, Golden Rule, The Gate is Narrow, Beware of False Prophets, By Their Fruits They Will Be Known, BE a Wise Builder, and When Jesus Finished the Crowds Were Astonished For, He Taught as One with Authority

Matthew 7: 7- 29

Ask and it will be given. The work of soul mastery and being a master facilitator of energy is to know that in this state of receptivity, of alignment and allowing, you can *ask and it will be given*. And then the work is to relax while the attracting forces organize in the field of all possibility to bring to you the highest and best of that which you are aligned with.

To attract a big dream requires that you believe big. It requires, according to the law, that you create a big vibrational momentum of Yes energy. An energetic momentum that has the capacity to catapult your ideas and possibilities into manifestation. This doesn't happen with a "yes, but!" That is why *the gate is narrow*!

Golden Rule. Do unto others as you would want done—regardless! Be the **Golden Rule** in action. Be willing to treat others as you want to be treated. This is expressing unconditional love. Unconditioned love is different from third-dimensional conditioned love, as is currently being lived. Conditioned love means that you must act a certain way, or I won't love you any more or respect you any longer. Conditioned love has perceptual boundaries holding it in place and you are happy as long the conditions are met. If and when they are not met, it makes you unhappy and you fight and argue for your perceptual limitations and boundaries to be reestablished.

The gate is narrow. The path is straight and narrow. Meaning that the higher the frequency, the less opportunity there is for wallowing in any type of negativity or show-stopper conversations. We're talking about the *yes, but; what if, if only, you don't understand, if only I was* . . . As your energy field rises in dimensional frequencies, manifestation becomes more instantaneous, therefore you must stay aware of what you are setting vibrationally into motion in every moment.

Ask yourself the question over and over, "Is this vibration I am holding serving me?" A simple question whose impact is far reaching.

Appreciate, appreciate, appreciate! Blessing confers energy upon a thing. An 1800s metaphysician, Emma Curtis Hopkins, said, "Demonstration is rest." Once you have demonstrated a spiritual principle in consciousness, sit back and get ready . . . Do not worry.

Wise builder. Be the *wise builder*, select a higher vibrational ground to build your spiritual foundation upon. Beware of those who tell you that you are Pollyannaish or that you just don't live in reality. Over time, they will see your life works.

By their fruits, they will be known . . . It is never "Why is this happening to me again?" You know why! Align with a higher energy and it will not happen again—or if it does, you won't care anymore because your perception has shifted. Perception has everything to do with how and what we will receive, so it is not about faking it 'til you make it! When you are faking something, what is the energy that underlies that thought? It is, "I am not this, but I will pretend."

Energy doesn't lie. It is better to find one thing you are good at and use that as your point of energetic appreciation than to try set up a field of fake-it-'til-you-make-it energy, for that sets up a subtle resistance because you know it is not true. Appreciate one good thing about who you are and allow that to attract more possibility. *I am a point of great possibility in this grand universe.* Your words have power, so be in appreciation. Ask, "Am I arguing for my limitations, or aligning to good through my words and actions?"

Why would you ever want to talk about what you don't want, when you know and understand that law responds to you as a vibrational point of awareness and attraction? Stop arguing for your limitations in order to be third-dimensionally right or justified.

> *Every word uttered energizes the ethers with a*
> *creative VIBRATIONAL impulse that in due season*
> *brings forth its image and likeness...*
> *Our bodies are weak or strong according*
> *to what we decreed for them...*
> *in accord with the obedient life and substance...*
> *The vitamins in thought are 1000 times more life giving.*
> Charles Fillmore, July 30, 1933

He taught as one having authority. Once the *Sermon on the Mount* was complete, an interesting thing happened, people began to realize that this man named, Jesus, taught as one having authority. In other words, metaphysically, Jesus was now living from a new level of energy, he had dominion over his words, command over his thinking faculty, and over his thoughts and emotions; he was living from a 4th dimensional state of conscious neutrality.

What this means for you is that your inner thoughts—which arise from your thinking faculty, your creative center, your inlet and outlet of ideas—are now recognizing a new power and energy at work—and you begin to align with that new energy. You begin to appreciate the new ideas brought forth from higher dimensional awareness, Christ consciousness.

They engage, support, and align with the new ideas, instead of fighting and resisting them, as they did in the past. Words spoken from our human mind and emotional conception are often weak. When you are aligned in Spirit, you hold an elevated energy and vibration, and you speak as one with authority and dominion.

Meaning your words are filled with a substance and energy that comes directly from Source, and people feel it. Words filled with the energy of Spirit are contagions and they have the capacity to effect change and transformation in your energy field. Are you experiencing a dominion over your words, and "command" over your thought and emotions? Have you fully released anger, jealousy, greed, the need to be recognized or in control, competitiveness, pettiness, playing small, trying to elicit sympathy, or getting people to feel sorry for you? Demonstration is the true index of this and of your spiritual growth.

There is a hidden spiritual practice, that the *Sermon* shares. It is concealed as energetic subtext and yet, it absolutely supports an accelerated movement of energy to the next level on the spiral, when engaged. As spiritual practice, ask yourself this question:

Is what I am about to say or do, aligned with the energy and vibration of non-judgement, non-attachment, and non-resistance?

If yes, move forward. If no, pause, breathe, go to your heart and connect with a higher energy. Appreciate something or someone.

The following is a short synopsis of how these three elements support your soul growth:

1. (Non)-Judgment
   a. Non-judgment—use faith rightly, not blindly—do not judge by appearance.
   b. Keep your eye single, and judge not lest you be judged. Karma is a bear!

2. (Non)-Attachment;
   c. Non-attachment—release old ways of being and thinking.
   d. Forgive and you will be forgiven. You can't serve two masters.

3. (Non)-Resistance
   a. Non-resistance—overcoming duality thinking and letting go of struggle, fight, and reactions.
   b. Take no thought. Be a non-anxious presence.

In summary, be the knowledge, information, light, and understanding of the *Sermon on the Mount*. Blessed are you who consciously live these attitudes (who are aligned with these fields of energy)! For wherever you go, you will confer a higher, lighter energy upon the energy fields in which you are interacting. For you now hold a dominant vibration that arises from a higher dimension of awareness, and energy doesn't lie!

Who you are vibrationally, determines the level of inspired action you can attract and act from. Will you attract as ideas for action from the guidance of higher awareness/Source Energy/Spirit? Or will you attract from your own human-survival-oriented ego? Each produces a different result in the unified field of consciousness and impacts the level of possibility and potential available to you in the moment. So, consciously carry the energy of knowing that this is a Yes universe. Know that the Universal Impulse is always attracting more to you, in alignment to your state of receptivity. Your perception has everything to do with how and what you are receiving, for it creates the state of receptivity for who you are!

Remember, your adverse ego-small self-little i, is the root cause of all your issues; you are our own devil tempting yourself to move into reaction. The energy of struggle creates chaos, which is a sure sign that you are not conscious in the moment, and the appearance of struggle came to teach you, about you. Struggle and chaos arise because we are not consciously practicing the attitudes set forth in the *Sermon*. The *Sermon* says, take the log out of your own eye first. You must recognize the false in order to know what is true.

When you are in alignment with Source you hold a state of awareness that is receptive to the multidimensional energies flowing without resistance and you discover that the time-space continuum collapses. The time-space continuum dissolves and the universal energies begin to work effortlessly on your behalf to attract the highest possibilities that you are aligned with in the unified field of consciousness; and synchro-divine events are now attracted to you, effortlessly.

Remember, moments of synchro-divinity are the result of being in vibrational alignment with Source Energy aligned. They are those unexpected and yet seemingly miraculous events that begin to emerge spontaneously within the field of your awareness. Unbidden, unexpected, and uncalled for, they impact you spiritually and shift your reality as they unfold. Synchro-divinity comes, to bring to you what you need, often before you had any idea that it was wanted. *Before they ask, I will answer.* Synchro-divinity comes to reveal mini miracles: right doors opening; right people showing up; providing information that shifts the trajectory of your life in that moment; and so much more.

The Energy Supporting Cana and Sacred Union

Cana is the place of the first "miracle" performed by Jesus and is described in John 2:1-11. The miracle, of turning water into wine, took place at a marriage, and its subsequent wedding-feast. Cana is a pivotal moment in continuing saga of soul evolution, for it represents another spiritual momentum point where many decide to turn back. They feel too much is expected of them. Metaphysically, here what the main characters and elements of this story means:

<u>Mary</u> is the ever-evolving feminine nature of the divine, in you. As your feminine nature evolves out of its 3rd dimensional unregenerate state of conditioned love and moves into higher 4th dimensional heart-awareness; it strengthens your innate intuitive capacity and ability to hear the voice of Spirit more clearly.

It is also, the role of the feminine nature of the divine to open you to a deeper understanding of detached compassion. Detached compassion tempers the intellect through the harmony of sacred union, wisdom and love in balance. It is divine, detached compassion that walks with you through the sacred fires of spiritual growth. Inviting you to continue to move forward, and yet, to be easy on yourself as you question, stumble, or even perhaps fall.

It is through the energy held in the sacred fires of initiation that the dross of sense consciousness is burned away, and consciousness is transformed into a healing balm of unconditioned love.

All the women portrayed in Jesus' life, are important elements of the Christ template. Each represents, in their own unique way, an aspect of the ever-unfolding and evolving regeneration of the divine feminine, in you and in the collective of humankind. Are you noticing how the feminine nature keeps showing up as an integral part of the spiritual journey?

<u>Cana</u>, a place of reeds, a measuring rod, rule, balance. In the body it is symbolized by the larynx. Cana of Galilee is the power center. How you choose to use your power, or not, is of great importance at

this point in the journey. This is the tipping-point place. Will you choose to move forward? Will you make the choice to allow Spirit-Source-I Am to move from being an energy guiding your life, to be the dominant energy guiding your life? Has your time come?

<u>Feast</u>, an appropriation in a large measure of divine potential, a laying hold of divine potentialities.

<u>Marriage</u>, the sacred union of two things and represents ideas and things coming into balance and harmonic resonance. Ideas like body/soul, masculine/feminine, thinking/feeling, and subconsciousness mind/Superconscious mind, coming together, working as one, in sacred union.

<u>Wedding</u>, a conscious union of energies, a sacred union of body and soul, thinking and feeling, etc. The merging of two into one, each retaining their unique properties, but in a state of balance those properties are utilized for the highest good of the whole, to support the whole.

<u>Water</u>, the great mass of thoughts that conform to the environment of old perceptions, and beliefs. They create pockets of energy, in accord with their belief, in the subconscious mind. This stops the vitality and energy of light and life from shining fully through the body system. Until spiritual awareness is demonstrated, thinking and feeling produce a very ordinary kind of existence (water), in this environment the vital energies (wine) can easily become depleted, and even run out.

<u>Wine</u> is the energy and vitality that forms the connecting link between soul and body. This energy and vitality must be present in large quantities before a union of the soul and body can occur.

In the Cana story, Mary, Jesus' mother, is the main character, as it is Mary, the feminine nature, who is actually invited to participate in the union ceremony. She in turn invites Jesus and Jesus, brings his disciples. At a certain point in the event, Mary (intuition) tells Jesus that the wedding-feast is running low on wine. Intuition, tells Jesus, that if he wants to fully participate, he better recognize that he is running low on the energy and vitality that forms the connecting link between soul and body.

He intuitively knows that this energy and vitality must be present in large quantities before the miracle of union can occur. Mary (intuition) then urges Jesus to express more of his divine potentialities, his Christ nature, and do something about the lack of wine (vitality).

Mary tells her son, it is time to stop playing around, get serious about the work you are doing. She tells him that it is time to transform his lower, water-like thought-energy, into a higher expression. Into the wine of vital, life-giving, higher dimensional energy. She intuitively tells him, it is time to consciously take another step, to initiate the sacred union process; to unite the subconscious mind and Superconscious mind, which will then allow Superconscious Mind-Spirit to be the dominant energy and guiding force in his life.

His intuition is urging him to do this now! Now, is time to demonstrate, not some time in future when you think you are more ready. Intuition is urging him to shift any remaining residue of lower emotions, doubts, and fears (water: the great mass of thoughts that conform to environment) into wine (higher dimensional levels of vitality and energy).

Even after hearing the inner Voice and knowing its power, Jesus still had his doubts about his ability to take the next step. He says to his intuitive nature (Mary), *"My time has not come."* In other words, he is saying, I am not ready, maybe I can't do this! Wait till later!

However, the intuitive self knows better and says to the servants (the thoughts that serve higher awareness), *do what he tells you* (ignore his doubts and support him in taking the next step, in doing what he knows he is ready to do).

Jesus immediately turns his lower, water-like, emotions, represented by the six water pots, the six nerve centers, filled with the water of life-nerve fluid, into a vital essence. He transforms them into the wine of "living" light and energy that is generated from the nerve substance. This is a pivotal moment, this moment of sacred union. Now the Superconscious mind, the I AM, takes dominion as the guiding principle of being.

"The Wine of life, or vitality must be available in large quantities before a blending of...soul and body (wedding-union), can be made successfully...new Christ life comes into a mind where old beliefs ...have been held and body is transformed..."[39]

The Energy Supporting Overturning of the Money Tables

This event comes after the Cana experience. It represents another layer of muck and density being transformed. The overturning of the tables of 3rd dimensional, mortal sense; throwing out the money

---

[39] Charles Fillmore, *Revealing Word*, Wine, Unity Publishing

changers; and the casting out of the thieves and robbers; all represent the energy of lower thoughts that rob you of higher dimensional awareness in the (body) temple. It is a conscious and necessary act, for only the "I Am" can live in a temple dedicated to God (higher awareness thought and emotion on all levels—body, mind, and soul).

The Energy Supporting Nicodemus & His Coming at Night

Nicodemus represents the unredeemed and unregenerate thoughts that often come to at night, in the darkness, and that try to shift you away from truth through an introduction of doubt and fear. Jesus shares with Nicodemus that in order not to be in fear, he must be born again. You must be born of life-giving and living energy of the Holy Spirit, the divine feminine.

Nicodemus didn't understand, and he goes away. This is another stopping point, a place where people jump off the spiritual path. Even after all the work that they have done, they allow doubts and fears to creep in and to take over, as the dominant energy instead of the I AM.

The Energy Supporting Healing the Centurion's Servant

This is another stretch moment. You are asked by someone to do something that you sense, for just a moment, might still be a bit beyond your capabilities. However, realizing the level of faith present, you gather your vital energy and send it forth with expectancy, faith, and authority—through the power of the Word—and the healing happens.

The Energy Supporting Anointing Lk 7:36-50 Mark 14

This parable is about forgiveness, spiritual discernment, humility, sacred service, spiritual mind discipline, higher levels of awareness, and engaging Christ consciousness. Everyone starts out with their feminine nature, their qualities of love, compassion, humility, sacred service, trust, surrender, transparency, authenticity, I AM dominion, etc., being in an unregenerate state. Being in a state of darkness and ignorance as to their true nature and often being unconsciously committed to the ways of sensation. These qualities must be lifted-up, regenerated, renewed, and reborn into a new light—as Jesus shared with Nicodemus.

A woman (representing the unregenerate, feminine nature, requiring purification—perhaps, Mary Magdalene), is visiting at Simon's house (a consciousness filled with intellectual knowledge, thus, not recognizing the qualities of love, service, and humility). She begins to wash Jesus' feet (cleanse her base of understanding with higher awareness). To do this, she utilizes a precious and costly ointment (love). Simon objects to this act, due to the expense of the ointment, and says the money could be better spent! Jesus immediately shares why this act is important to the spiritual journey.

Jesus lets Simon know that this act of pure and unconditioned, humility, love, surrender, and compassion, is another phase of releasing and letting go of conditioned reactions. It is an act that must occur in order that you might prove that you are in service to something greater than your self. It is a choice point moment, where love is regenerated and restored to its true higher nature and awareness. From this moment on, unconditioned love and the feminine nature is fully in service to Christ consciousness and to the greater good of humankind. Remember, every character and element found in the Christ template, is an aspect of you, revealing what you are to either ready to release, or take on in fuller expression.

The following is a quote from my book, *Embracing the Feminine Nature of the Divine*. It is from a lecture by Charles Fillmore, given in January 1916, on *The Authority of Divine Feminine Restored*. It one of several series of lectures that Fillmore gives on the role of the divine feminine in regeneration and spiritual evolution. He initiated these lectures, after he returned from being away for several months, due to a serious health challenge. This health challenge, and subsequent healing experience, appeared to awake in him higher levels of dimensional awareness regarding the role of the divine feminine in spiritual evolution. These new teachings were levels beyond the introductory mother/father God awareness and teachings Charles and Myrtle began with. In this lecture his message speaks to the metaphysical role of Mary Magdalene as a character in the life of Jesus.

...We know that "man" is masculine and feminine in nature...We are in our present civilization coming to a place where [all] must recognize the Divine Feminine...

What is the true relation existing between the masculine and the feminine in the doctrine of ... Christ; in pure Christianity...in esoteric Christianity... [Jesus] said; "Know ye not that he created you in his image and likeness, male and female"...this means that there is a divine marriage...a marriage of Spirit; there is a union...the masculine and feminine... qualities which exist in being...

Mary [Magdalene]...was his Divine Feminine in personal manifestation... She represents the Divine Feminine which required purification...the Divine Feminine must be brought into expression.

The Energy Supporting Mary Magdalene and the Seven Demons

For the I AM and its twelve supporting faculties to minister successfully within the organism, they need the sustaining assistance of the feminine quality of love. Love, which has been regenerated, lifted-up, and redeemed of its demons and thus, transformed from a mortal, conditioned expression into a unconditioned, spiritual expression. Mary Magdalene represents that redeemed and regenerated quality of unconditioned love.

The Energy Supporting Transfiguration

After a certain amount of spiritual growth through sacred service and demonstration, you will experience a spiritual transfiguration. Transfiguration is a supernatural change of appearance that takes place as you experience the full flow of divine power, luminous energy, and light moving throughout your being.

As you continue in your acts of meditation, the holding of elevated thought and emotions, and the embodying of the 4th dimensional state of awareness of, there will come a time when you will experience a rapid radiation of luminous energy. This luminous and radiant energy will cause a dazzling light energy radiation from all parts of your body. You will look and feel like you are afire, with a luminous, radiant light!

This experience occurs because you have established through demonstration and service the dominion of the facilities of faith, love, and judgment (spiritual discernment) within your body system. These are the three major facilities that support you in the spiritual and emotional maturing of your adverse ego and in bringing forth a greater expression and awareness of Christ consciousness.

However, as with any experience, you cannot pitch a tent, and stay in this high place. You must at some time return to everyday life and put your newly transformed belief to work in your life.

5th Initiation: Gethsemane

Through through your matured faculty of judgement, which now holds the energy of spiritual discernment, intuition, divine will, and the capacity to call forth of the energy and fire of Spirit, you now make decisions and choices in new way. You observe, evaluate, compare, pray, and discern and then make your choices.

Spiritual discernment can only occur through righteous thinking, which is a combination of right-thinking and right-understanding, aligned with higher dimensional awareness. Right understanding comprehends the realm of higher ideas through spiritual guidance and wisdom, and not through the intellect, alone.

Right-understanding also brings divine will into play, as a partner. Will is your executive power and faculty of mind. Discernment and understanding weigh the information differences between your choices—will determines which choice you will choose and the subsequent action you will take. Through the power of will, along with discernment-understanding, you determine whether your choices will be aligned with the adverse ego, little self, or with Spirit. You will what comes to pass in your experiences through your choices and decisions, and the resultant actions taken because of those choices.

It is in the Gethsemane initiation that you truly discern and establish in consciousness the idea that all decisions–regardless of consequences—must be made in alignment with the divine will of Source-Spirit-God-Good. That regardless of what you personally want, it is imperative that you discern and evaluate all situations and conditions and then make the conscious choice and decision to not to judge by appearance or sense perception. You decide that you will make all choices using the matured faculties of spiritual discernment and divine will. Gethsemane is also the place where Jesus has to make a major judgment call about his future, his destiny, his "cup."

The Energy Supporting Why You Can't Deny on the Cup!

Gethsemane begins with Jesus praying and sweating blood, while the disciples who are supposed to be holding the high watch with him, fall asleep. The sweating of blood, metaphysically, means that he is in deep concern and consternation over what is to take place next in his life. In fact, he is asking to have the "cup" of destiny, taken from him. Metaphysically, the cup is his purpose and destiny, the crucifixion.

Your "cup" will be different, but the feeling in the midst of the situation, will be the same. You will "sweat blood," and be in deep struggle over what is to come or, might come next. You will yearn for friends to support you, but in the end, the eyes of your friends will be heavy with sleep. You will be alone in the making of your decisions. Friends, lovers, parents, sisters, or brothers cannot enter the heart of your struggle, you must face your own personal "cup" situation, alone.

Your specific cup situation or condition is part of your soul journey, your personal evolutionary process. It will come in any variety of forms—illness, loss of loved one, loss of something important to you— and more. You, karmically, cannot afford to let it pass, for it is important to be tested in every area of mind. To overcome this initiation is to be able know so deeply, who and what supports and sustains you.

This is a moment where your faith must be so strong, that you can say to the Pontius Pilot's that rise up in your life — *"You could have no power over me, unless it was given to thee by God."*

Gethsemane is important for it prepares the way for what is to come, the crucifixion. In fact, Gethsemane makes the crucifixion possible. Many take the first four initiations successfully; however, only a few have the courage and strength of purpose to pass the severe ordeal of Gethsemane. And therefore, appear to fail in their attainment of the next step of spiritual awareness and attainment in consciousness.

The Energy Supporting "Nevertheless, Not My Will but Thy Will Be Done."

A full surrender of the little self—self-will—adverse ego—occurs in Gethsemane. It is a key to the development of the regenerated consciousness that reveals the fullness of Christ. And the act of surrender comes through one of the most practical spiritual tools that will ever be shared with you: *Nevertheless, I Am Willing!*

Even when you are afraid, even when you don't want to, even when it seems to hard, or to painful, you will set aside your will and your personal preferences and do what is right. You will do the highest and best of all involved. Nevertheless, I am willing to move foreword. Nevertheless, I am willing to forgive. Nevertheless, I am willing to do what I don't want to do. Nevertheless, I am willing.

Nevertheless, I am willing to merge with the Divine, to release all claims of fear, and do the will of Spirit. The fears and claims around not getting what you want, how you want it, and when you want it are being transmuted, as the will of the small self, the adverse ego-self is being translated into Divine Will, higher heart-wisdom.

Being willing to surrender into Source Energy/Divine-Will is a call that can only be made by the soul. It is a permission granting moment. Source will not do anything without first, having your consent for change. Thus, the choice to align vibrationally with the will of Higher-God-Source-Spirit must be made by the soul, before you can advance to the crucifixion experience; which is the segway into a resurrection of consciousness. The crucifixion experience is an alignment with the vibration of truth, that makes possible the resurrection into the vibration of freedom and liberation. For alignment with higher dimensional awareness, transmutes the energy of resistance and opposition, and thus, it is dissolved, released, and surrendered.

However, first, the claim of being, must be an alignment with the vibration of being willing—*Thy Will be done*—or, you will not succeed in the resurrection of consciousness into a higher vibration of freedom and liberation. Which is a state of awareness that knows that it is free, liberated, and released from the resistance, opposition, and false claims of the small self, the adverse ego; free from the energy of the false claims of addiction, power, greed, ego-tricks, needs, etc.

Collective consciousness is transformed, as each individual consciousness, one by one, is willing to claim first, the vibration of being willing. Being willing to consciously surrender one's personal will in order to merge into and align with the vibration of truth, which is the will of Spirit; and finally, then to be resurrected into the full vibration of freedom, in higher dimensional awareness. When enough do this, a critical mass will be met, old timelines and paradigms will collapse, and higher dimensional awareness, Christ consciousness, will be assumed by the collective.

The Energy Supporting Judas and Betrayal

When circumstances feel intolerable, due to betrayal, be it by another person, your body, an illness, your job, etc., it usually means you have outgrown a situation and are ready to change. The question is will you do it with ease and grace or, through resistance and struggle?

Betrayal is a heart-opening or heart-closing experience, it is your choice. It is in Gethsemane where friends forget about you, where friends who you always thought would stand by you, who swore they would stand by you, fall asleep, leave you, cheat you, cheat on you, and ultimately betray you. On the journey, at some point, the body will betray you.

Regardless, of the form the lesson takes, all experiences only come to teach you about, who you are and who you are not! In Gethsemane you learn, through betrayal, that there is nothing in life to count on, but Spirit, One Presence, One Power, active in your life.

The whole Gethsemane experience is about surrender and self-renunciation, moving from my will to divine will. No matter what the betrayal or call for surrender looks like. Judas' role is in Gethsemane appears to be one of betrayal, however, he plays an important part in the supporting of Jesus in his moving forward towards his soul purpose. Jesus couldn't have done completed his mission, his soul purpose, without Judas.

Regardless of appearances, at this point of human evolution, betrayal is an imperative learning tool, as a part of spiritual growth. It often takes a betrayal to wake you up to what is imperative and necessary for the step of commitment and for your spiritual journey.

The Energy Supporting True Power and Pontius Pilate

Jesus is taken in custody after Judas' betrayal, and Peter's denial of knowing him. While in custody, Pontius Pilate, the governor, questions him. Jesus shows no resistance, he doesn't try to defend himself, or fight for his freedom. Jesus only responds to Pilate with these words (a paraphrase), "It is as you say. For you could have no power over me, unless it was given to thee by God."

Whatever appearance of adversary or betrayal you face—divorce, death of a loved one, cancer, body betrayal of any type—it can have no power over you, unless you give it power, unless you choose to make fear of it, a power. True power resides in Source Energy. Stay focused, centered, and aligned in the vibration of truth, and truth will prevail!

6th Initiation: Crucifixion

The 6th initiation is a crucifixion experience, it is another heart- opening experience on the journey to realization and acceptance of self, and as Self-I Am-Christ.

The Energy Supporting Collective Consciousness

The human race has formed laws of physical birth and death ...the sum total of these laws forms a race consciousness separate from and independent of creative Mind. When creative Mind sought to help... spiritually, the mind of the flesh opposed it... the great need of the human family is mind training [spiritual mind discipline]. *Jesus showed us that mastery.*
Charles Fillmore

The Energy Supporting, I AM Dominion

The I AM vibration assumes full dominion through the process of crucifixion. The process of claiming and owning your I AM, Creator Energy, as your full identity creates such a major shift in you that ultimately, it affects collective consciousness, as a whole.

Crucifixion metaphysically means, the process of crossing out the last vestiges of old beliefs and perceptions holding you hostage to lower dimensional vibrations. It is a process that energetically supports the full maturation of the adverse ego, in order that the energies of unconditioned love and harmonic resonance can do their work in the resurrection process. Although, the experience of crucifixion does not appear as such, its work—as you cross out, dissolve, transform, lift-up, overcome, and transmute the energy of negative-error thought and emotion—creates space for higher dimensional awareness to fill you.

The Energy Supporting the Outer Garments

The "*outer garments* " which the soldiers divided among themselves represent the various energetic-clothing and energetic expressions that your mind clothes itself with, around the ideas and teachings of the concept called God. You then take those concepts and mold them into beliefs–called doctrines and creeds. You then choose certain elements of the various doctrines or creeds to suit your purpose, and all of this, supported you in the creation of your intellectual conception and perception of what you know call your religion, spirituality, and/or truth.

Jesus did not need his clothes—his old beliefs and perceptions—to go to the cross and you too must give up the protection that your outer garments of thought-energy provides. They are of no value now, for there are no limited concepts of doctrine or so-called "truth" in Christ consciousness.

The Energy Supporting the Inner Garment-Seamless Robe

It is said that Jesus' inner garment, a seamless robe, was gambled for, to determine whose it shall be. Therefore, the soldiers say: *"Let us not rend it, but cast lots, whose it shall be."* (Jn. 19:24). Its value was recognized.

The seamless robe represents the the inner-thought garment of mind, which holds an energy of higher dimensional thought that is woven without any breaks in the high realizations of Truth. The seamless robe is interlaced or woven with the energy and substance of Christ, as the spiritualized body and body consciousness, as One. It is an interior spiritual understanding of truth and life as eternal. It is known without a doubt.

Life is eternal, unchangeable, and unalterable in its divine pattern of wholeness. Thus, it is unrendable, unable to be torn in two, unable to return to lower energy of duality thinking.

The Energy Supporting the Role of the Two Thieves

The two thieves or robbers energetically represent those parts of you that must still be uplifted, transformed and overcome. The energy of these two thieves, as long as they are present, rob you of the capacity to take on the full realization of Christ. The energy of these two thieves represent:

1. Any remnant of the belief in matter or, the energies of 3rd dimensional reality as your reality *(right-hand side of the cross)*

2. The illusion of duality or, any form of "evil" as reality *(left hand side of the cross).*

Right-hand thief or robber (belief in matter) is salvageable, for he is crucified and saved at the same time. It takes effort to overcome materiality, to discern the essence of the energy of higher dimensional awareness-spirituality from the remnants of your formed false concepts and beliefs. This thief repented of energy of sense evidence and opened the way for the material body to transform into the energy and substance of the spiritual body. The robber element disappears, and substance-energy remains. This too, is you in that moment when things seem darkest, and you start to have an aha. And the aha, gives you the glimmer of belief-faith-hope that knows you can move forward in a new way, in a new energy.

Left-hand thief or robber (Latin. Left-Sinister) has no salvageable substance, he represents the firm belief in duality, in good and evil – he is steeped in the energetic illusion of two opposites. And what is true is that all of the false supports of the human mind such as personality and the adverse ego, must be dissolved, transformed, transmuted, and/or released.

A mixed state of mind cannot produce
Pure, Absolute Perfection.
Charles Fillmore

The Energy Supporting the Cross

The cross is a symbol of the energetic-forces in you being adjusted into a right relationship. Here you mature the lower energies of personality in order that the Christ mind may be more fully expressed. The cross, and the sense of being on the cross, represents a time of crossing out, releasing dense energy, and spiritually maturing those states of consciousness, known as mortal mind. This includes the spiritual maturing of all false beliefs, false perceptions, illusions, veils, shadow aspects, etc., all of which burdens the body with the dense energy of false beliefs.

At the center-point of the cross, the heart of the cross, is where the action takes place. It is in the center-point of the heart that the final overcoming of any resistance that remains towards a situation, person, condition, takes place. The I AM is established as the one and only power and energy working in your life. One Presence and One Power, active in your life, that absolute knowing calls forth the fullness of the Christ consciousness, and it brings with it the capacity and ability to move forward in spiritual resurrection.

The Energy Supporting "Eli, Eli, Eli, Lama A'sabathani"

This quote is considered a part of the last seven words spoken by Jesus on the cross. Interestingly, it is found in all four Gospels. It has historically been interpreted as meaning, *"My God, My God why has Thou forsaken me?"*

Was this a moment in which he lowered his energy energy field and dropped back into 3rd dimensional projection? Or, was it something else? A misinterpretation, perhaps?

Modern day Aramaic scholars are now setting forth a new premise, through a new translation of the statement which now reads: *"For this I was born, for this is my purpose."* This creates a whole new level of energy and meaning for these words and for the experience. For this I was born, for this is my purpose.

It makes perfect sense in light of his Gethsemane experience. In Gethsemane he does, for a moment, lower his energy field, and laments over the "cup"—his destiny, it was a moment of "humanness." He sweat blood over it, but he knew his mission and his soul purpose. And he has he divine realization and lifts his energy field by knowing—nevertheless I am willing—on every energetic level—I am willing.

He is willing to complete his mission, his soul purpose, his soul theme, and his soul contract. He is doing it, not for himself, but as an act of sacred service for humankind, and all sentient beings, he is accomplishing what he had set out to accomplish, what he came to do.

If you are moved to do so, close your eyes, and repeat the sacred words of, *Eli, Eli, Eli, Asabathani (A-lee, A-lee, A-lee, Ah-sab-ah-tah-ni).* Repeat them over and over like a chant, for several minutes. Allow them to permeate your being and cellular structure, feel the words doing this. As you do this, remember the underlying idea of what they mean, *for this I was born, for this is my destiny.* After a few minutes, be still and observe what arises.

The Energy Supporting the Appearance of Death

During the crucifixion experience all remaining remnants of the adverse ego, of lower 3rd dimensional energies, are transformed. They are consciously, spiritually matured, in those final moments. You do not ever kill off anything in consciousness; it dissolves and disappears from lack of substance to sustain it.

What this means from an energetic standpoint is that while the body appears to be dead, it is not destroyed, it is redeemed, energetically transmuted, and spiritualized through the act of overcoming the appearance of the death of old energies. Remember, the crucifixion is necessary for it makes possible the transcendent glories and energy of the resurrection. When the appearance of ego-death occurs, it is a final step in the overcoming of the issue/s that betrayed you. Remember, crucifixion is not a onetime thing, it happens over and over as you evolve spiritually. It will happen every time you let go of layers of density, karmic muck, and shadow material.

The crucifixion experience is a part of the releasing process of the "lower-level energetic beliefs, perceptions, and ideas" that have held you hostage. The releasing of those beliefs invites a transformation into a higher level of awareness and power, and you are imbued with a new energy.

> *Our word has only the life and power*
> *of the idea that animates it.*
> Georgiana Tree West

The Energy Supporting the Veil of the Temple Being Rent

In that final moment, when the appearance of the last breath is taken and what looks like death appears, the earth shakes, and the veil of the temple is rent. The metaphysical energy underpinning this is that of being at what appears to be a breaking point—and you do—for at this point, you energetically break through the veil. The veil represents the energetic illusions that have held your false beliefs in place.

Now, because the veil is rent, or torn in two, you can see God face-to-face, with clarity. Clearly, you see the gift in the situation, you have your answer, or your ah-ha arises for there is no more "stuff" hidden in the deep recesses of subconscious mind to obliterate the view of face of God

At the moment when the veil is rent, the earth (your body and being) is shaken also. This is an indication that a major breakthrough is happening within your being, systemically. As your world is being rocked, you are being transformed. Through this shake up of earth the dead (your bound thoughts) are energetically set free. This journey of soul mastery is not easy. Is it any wonder, so many fall by the way side?

Human mind (the intellect and its companion, the adverse ego, the small self) is ever hostile to the higher-level energy, vibrations, and frequencies that hold 5th dimensional awareness and Christ consciousness. Jesus began his journey as the *son of man*— as all are in their human nature—and it is the only son of man that can die. He grew into the *Son of God,* a level of consciousness that knows life is eternal.

Jesus was not a substitute for you, he was the way-shower, and a prototype. He was the prototype for the conscious creation of the Christ template. His experiences and actions left a template that you now can use to advance your soul growth. For the template reveals a step-by-step, instruction on how to come into the fullness of 5th dimensional awareness, how to be master facilitator of energy, and how to live from a consciousness of soul mastery.

7th *Initiation. Resurrection and Ascension*

The Energy Supporting the Tomb & Regeneration

After what appears as "death" you are then taken into the tomb, a place of silence and of waiting. You wait for the full energetic and light-body maturation process to complete itself—it cannot be rushed—it takes three days. This light-body energy maturation process prepares the way for resurrection and ascension. Mr. Fillmore shares this regarding the *Tomb* experience:

The appearance of a human body went into to the tomb, and came out transformed, forever, into the spiritual body–FREE—with consciousness transformed, purified, regenerated, illuminated and reflecting only the Light of Christ. It is only our conscious awareness of duality that is dead, not our Christ Nature, for it is eternal and unchanging and always within us.

The transformation that took place in the body of Jesus was accomplished through the spiritualization of his mind and that brought forth the spiritualization of His body and ultimately the world He lived in. "The tomb where Jesus was laid to rest represents an elevated, peaceful state of consciousness in which He rested the three days previous to His resurrection. The word of Truth within Jesus did not die but was quietly spreading from point to point during this period, getting ready for the supreme test: the overcoming of the appearance of death.

For us, the tomb represents a high state of consciousness in us in which we improve in character along all lines. We not only grow into a broader understanding but also, we increase in vitality and substance. We are resting in God, and at the same time gathering strength for the power of greater demonstrations to follow... This process continues until the whole consciousness is vitalized by the Holy Spirit."

The three days in the tomb represent three steps in overcoming error. First, nonresistant and humility; second, the taking on of divine activity, or accepting the will of God; third, the assimilation and fulfillment of Divine Will.

If a human/mortal/carnal/material consciousness has given us our bodies, with this material appearance and its properties, then a transformed consciousness must culminate in a transformed body by natural law. This is the secret and mystery of regeneration and the mystery disappears, when the secret is found.

Crucifixion is transformation through demonstration, through mastery of energy, through the power of regeneration. Regeneration is the capacity to renew the mind over and over, until you bring all the life forces of your being—spirit, soul, and body—under the auspices of the Superconscious mind. This work eventually creates a change in the DNA, and the cellular structure of your body, calling forth more light.

Regeneration, a change in which abundant spiritual life, even eternal life, is incorporated into the body. The transformation that takes place through bringing all the forces of mind and body to the support of the Christ ideal. The unification of Spirit, soul, and body in spiritual oneness...These over-comers shall make a new world, a new heaven and a new earth... the world is waiting for the manifestations of the Sons of God." [40]

The Energy Supporting a New Jerusalem

Having experienced the seven initiations of soul evolution and mastery and having dissolved the illusion of "sin" and duality thinking, you have energetically "died" to them. You have resurrected into a new awareness, and you now stand at the threshold of ascension into a new life, a new energy, into the *"New Jerusalem."*

The New Jerusalem represents a 5th dimensional (and beyond) spiritual consciousness that has united the forces of spirit, soul, body and mind under the auspices of the energy of Christ consciousness, unconditioned love, life eternal, and eternal peace. Love, life, and peace at this level of awareness is steeped in spiritual understanding and requires no outer protesting or fighting to make it real, it in itself is the alchemical transformer of energy.

Right where you are God Is, and all is well. All is well, for now wherever you go, whatever you encounter, you bring the energy of Christ-Source with you. And by virtue of higher dimensional energy and awareness being present, in your presence, energy in the space around you transforms.

---

[40] Charles Fillmore, *Revealing Word*, Regeneration, Unity Publishing

6

Mastering Abundance

Everything is energy and that's all there is to it.
Match the frequency of the reality you want and
you cannot help but get that reality.
It can be no other way!
Unknown

The last of the quadrilogy to be presented, as a part of the soul mastery wisdom-teachings, is how-to energetically master abundance. If you have been consciously participating with the principles, energy, and application of the last three chapters, then by now, mastering abundance should come easy to you. Here is my story of how the demonstration and mastery of abundance came about, as a living reality in my life and consciousness.

It was always at night that I would fly. From the age of four to seven years old, the experiences of flying occurred, and vivid memories of these experiences still live within me. I can still feel it cellularly when I think about that time in my life. I can feel what it felt like to be weightless and swooping effortlessly around the old three-story house. In the house, out of the house, up the stairs, down the stairs, watching my loved one's sleep, and so much more, often waking up at the bottom of the stairs on the first floor—not knowing for a moment, how I got there.

At times, my flights would take me outside. I would stay close to my family home, but always I would swoop down and run my fingers through the grass and, lo and behold, the grass would turn to gold coins. I would effortlessly take flight among the trees and reach my hand out to touch the leaves, and the leaves would turn into dollar bills. Money, money, money. I was Master of It All and it was mine to use as I wanted—so I felt.

The feeling I had as child was not so much about the wonder of flying and taking flight as it was that I knew that I was rich, that I lived in an abundant world, with everything appearing luminous and having a radiant, glowing quality to it. Anything I wanted could be mine. My soul knew I had lavish abundance at my command.

It was my secret world and I shared that secret world with no one. I can still elicit the feeling of joy I felt. It was breathtaking. Then, at the age of six-and-a-half—and this memory, too, is vivid; it is burnt into my neural pathways—I had my first memory of a choice-point moment.

The choice I made was to share my secret with my uncle. He was my favorite person in the world. I trusted him more than anyone. I recall crawling up onto his lap and telling him that I had a secret to share with him. He lovingly looked at me with his big blue eyes and listened with rapt attention as I wove my tale of wonder, awe, abundance, riches, flying, and more. I remember how safe and warm I felt in arms. Most of all, I remember how excited I was to finally share my secret with someone.

When I was finished describing the awe and mystery that had revealed itself to me over the past few years and expressing how so abundantly rich and wealthy I felt—in a little child's words, of course— my sweet darling uncle looked at me with those big, intense blue eyes, seemingly into my soul, and he said:

"Toni, you are <u>not</u> rich. We are <u>not</u> rich. Honey look at your clothes. We live in Frog Hollow, for God's sake. We are poor."

Boom! Really, a feeling like a sonic boom went off inside me. For the first time ever, I looked around our house—really looked. I looked at my tattered pajamas, the worn tables and chairs, the lumpy sofa, and the chipped paint on the walls. And for the first time, I knew what poor felt like. From that moment on, my dreams of abundance eluded me, and I lived my life for the next twenty-three years believing I was poor, that there was never enough, and, most importantly, because there was not enough, I was not enough.

It showed up in all I did, in how I felt about myself, and how I expressed myself in and to the world. It became my vivid, vibrant emotion, memory, perception, and belief. It was what I emotionally and vibrationally broadcasted into the world through anger and aloofness. And most importantly, it became the reality of the landscape of my life. I was miserable and I expected misery. People would say to me, "You are attractive, why don't you ever smile?"

I remember thinking, "For God's sake, what is there to smile about?" And that is how reality revealed itself to me through the experiences of my life, until I awakened to the truth of my being and to the knowledge that this is an ever-abundant world.

First and foremost, know that nature's natural state is abundance. Understand this, and you begin to shift into a new awareness of the abundance that surrounds you and lives within you.

You are the master of your masterpiece!
You have the capacity to repaint the landscape
of your abundance reality!

The potential for abundance to express in all areas of your life, is not just a promise, it is a living reality that is encoded into your energetic field. You just have to learn how to access it. How do you do this? First claim, unwaveringly, the Truth that is already encoded in your energy field — *abundance is my natural state.*

From an energetic and spiritual perspective, when you have established neural pathways for abundance, you hold the following ideas, tenets, and premises as an absolute—ab-SOUL-ute—knowing:

- Abundance is your natural state of being.
- You are a vibrational being living in a vibrational universe.
- Your vibration holds a frequency, a frequency governed by law.
- You create your reality, experiences, and manifestations through and in alignment with the vibrational frequency you hold and emit—it is the law.
- If you hold back, it is your lack.
- All things hold an energy and represent a frequency level—this includes your money, all tangible goods, trees, flowers, thoughts, feelings, emotions, goods, etc.
- Be the difference and you will make a difference — be conscious neutrality in action.
- Regardless of what you choose to share or the amount you choose to share, share from a vibration that holds a generosity of spirit and a consciously connected heart—a heart aligned with a higher mind consciousness and dimensional vibrational frequency of the principle of abundance (no trace of lack).
- Honest self-observation of your emotions, thoughts, words, and who you are in every moment is the training tool that opens your heart toward an understanding of what sharing from a generosity of spirit means: an appreciation for all things great and small and for the expansion of consciousness.
- Mastery of the spiritual principle of abundance causes that principle (or any principle) to be a demonstrable, sustainable, and repeatable pattern that is forever at your disposable.

- Participating in the development of that sustainable pattern creates a shift not only in you, but also in and for collective consciousness, the greater good of all.
- Mastering abundance is not about personal gain or what you can get. Its underlying intention is to share from a consciousness that knows that all that is being shared is in service to your soul growth and the raising of collective consciousness toward a greater good. It starts with self; it starts with you.
- Consciously working the elements, tenets, and premises that underpin mastering abundance changes your vibration and vibrationally signals the universe/Source Energy/the Unified Field of Consciousness to initiate support to match your new level of vibration regarding abundance.
- Mastering abundance is understanding, with clarity, that as you consciously master abundance you are creating a mindset that plays an imperative part in the unfolding of the cosmic blueprint for the collective, evolutionary advance of humankind. It is a personal soul-work done in service to the greater good of all: the conscious uplifting of collective consciousness.

Whether you share money, time, talents, knowledge, joy, wisdom, higher dimensional awareness, etc.—you cannot do anything for selfish purposes—nor, can you hoard. Both your soul growth and your personal gains must be shared! As you choose to shift your awareness of what abundance means to you, you clearly hold an intention that you are not doing this work just for yourself. As you do the work, which must be done individually, you do it knowing that it is always for the greater good of all sentient beings and the collective good of all. This may seem like a small thing, but it is huge when held consciously as part of a conscious outcome.

Prosperity, when worked from a third-dimensional mindset and perspective, is usually oriented toward survival and selfishness. What can I give to get, to manifest for me? What can I get for me in order to increase my wealth, make me more comfortable, rich, powerful, etc.? Whereas, when working from a higher dimensional awareness (remember, dimension means a level of new information), one that is oriented toward the demonstration of the spiritual principle abundance aligned with and oriented toward a consciousness of sharing that supports the collective good of all, you begin to consciously understand

and overcome the needs of the third-dimensional greedy-selfish-survival-me-me-oriented ego.

Humankind, as a species, has been mesmerized into forgetting how ever-abundant life can be. Collective humanity has been lulled into a sense of playing small through a belief in scarcity. The time to pierce the balloon of false perception and to release that which has been previously and collectively agreed to is here now!

And, once you have participated in, even just a little soul mastery work, you will start to observe that the unified field of consciousness will initiate a partnership with you. Synchro-divine events (synchronicities) will begin to occur unexpectedly, over and over. The unified field of consciousness begins to align with you and draws to you that which you require, almost before you even ask, in order to support the furthering of your soul work in the world. A work which is always oriented toward the greater good for all sentient beings. Regardless of the title your work holds!

It is time to be in remembrance of the truth of your abundance consciousness, and how to do this has been shared over the millennia by many master teachers. One of these master teachers was the man known as Jesus. As I began to grow in spiritual maturity, I began to release the stream of resistance and negativity I held toward my religious past. Suddenly, Jesus' teachings began to come alive for me in a new way.

When you are ready for a shift in energy, information comes to you. But it only comes as you are prepared to receive it. You don't have to go get it—it comes to you! This may seem strange, but it is true.

Having prepared yourself, through study, classes, etc., the unified field of consciousness—the field that is in interconnection with all things, ideas, and persons—knows when you are ripe and ready for new wisdom and It attracts it to you. That is at least how the teachings and messages have unfolded for me on my spiritual journey, and it is how mastering abundance became a living energetic reality for me.

As a teaching, mastering abundance is focused not on money or getting things, but on how to demonstrate spiritual principles (e.g., abundance, life, joy, peace) as a consistent, repeatable, and sustainable activity in mind and consciousness.

The idea of the unified field of consciousness and all possibility residing in that field, has been mentioned several times for it is important in the context of mastering abundance. In quantum science, it says all possibility and potential exists right here, right now, in the unified field. You just have to align with the level of possibility that you desire.

From a metaphysical, mystical perspective, this concept was set forth in the Hebrew testament through the words found in Genesis 1. *"In the beginning, God created the heavens and earth."* Mystically and metaphysically interpreted, this quote says:

> For forever and always (*in the beginning*), Universal Energy (*God*) created the potential for all levels of possibility, from the highest to the lowest (*the heavens and earth*). And this potential waits for you to activate a possibility through the conscious sharing of your word, of your thought-feeling vibration. Through the speaking your word aligned with a higher vibrational awareness and power *(Let there BE Light)*.

If all possibility exists now, there is nothing to "get." You only have to align vibrationally with the possibility you desire, and that vibrational match will draw its equivalent to you. This means it is not about picturing what you want or affirming what you want—it is about knowing beyond a shadow of a doubt that you are aligned with the highest possibility of abundance available.

That sense of knowing is a vibrational signal that is a qualifying energy for that which you desire. It will draw back to you from the unlimited field of possibilities, all that you are in alignment with, vibrationally. As you do this, you are consciously engaging the practice of being an arc of infinite abundance in expression, a point of attraction consciously causing effect!

Remember, the universe doesn't respond to what you say — but to what you mean. And meaning is determined by what you claim, feel, and hold as a knowing. Knowing is a vibration, created from thoughts and feelings!

You can say all you want that you are abundant, rich, or at peace, but if—you don't feel it, you don't know it—and thus, you are vibrationally out of alignment and it cannot manifest. You must be vibrationally aligned in consciousness on all levels of thinking, feeling, knowing, sensing, and acting to cause effect. Energy doesn't lie!

Additionally, as you transform yourself, you transform the collective conscious. So, what you think, feel, and know about yourself and abundance matters. You are here to make a difference, not only for yourself but for all sentient beings. The vibration you hold makes that difference!

The subconscious mind EXPECTS
to RECEIVE from Source ... not outer channels...
Charles Fillmore

The practice of tithing is thousands of years old and is underpinned by centuries of third-dimensional constructs and understandings as an expression of exchange. It is older than Mosaic Law by hundreds, if not thousands, of years, as it was originally a pagan-oriented system that was initiated by using vegetables as an exchange, not money.

The first tithe of goods and monies is found early in the Hebrew scriptures and it was given to Melchizedek by Abraham. When Melchizedek saw that Abraham had returned safely and as a victor from a major battle, Melchizedek asked if he could bless him. Abraham was moved by that conferring of energy. He obviously felt the activity and energy of God action in that blessing, because he willingly offered to share ten percent of his spoils with Melchizedek. There was no request or sense of obligation imposed.

However, later in scripture there is a shift that occurs in the act of tithing. It is imposed upon people from a mindset that is more about giving of money or goods to sustain the church, priest, landowner, overseer, or landlord. Thus, it is now both imposed and considered an obligation, a tax, or dues for being allowed to stay or roam free in a county, region, country; to participate in a religious system; or to curry favors.

So, there was instilled in the mindset of a tithe a giving to get! Which definitely is still alive and well now— think of *The Secret* and consider how most prosperity programs' basic premise is "give to get" money, houses, cars, etc. Thus, the word tithe carries third-dimensional energetic baggage regarding giving. And whether you see it or feel it, that baggage underpins the energy of tithing and giving until you consciously shift it.

Charles Fillmore says, "*If you think lack, you are connecting with all the other people in the world who are thinking lack also.*" The same energy is attached to the word's "tithe" and "give," because of how they have been defined for many centuries.

When you unconsciously connect with those words, you are connecting with the consciousness of all of those persons who have, over thousands of years, tithed or given out of a sense of conditioned obligation. Thus, unconsciously, you are connecting with same sense of obligation or desire to-get something-for having given.

So how can you transition from a third dimensional sense of tithing, to living from an understanding of what it means to master abundance and to be an arc of infinite abundance in expression? You simply start by pausing, taking a centering breath that anchors your energy in the heart, and from that field of resonance, create a conscious intention. *I am one with the wisdom encoded in the principle of abundance!* Then go on a quest of discovery. Here are two questions to support your quest:

- What does it metaphysically mean to be an arc of infinite abundance in expression?
- What is the metaphysical meaning of Malachi 3:10, as an expression of abundance?

1. What does it metaphysically mean to be an arc of infinite abundance in expression?

Defined metaphysically, an arc is a sustained, luminous, energetic discharge between two points. Thus, to be an arc of infinite abundance in expression would mean that you consciously hold in consciousness the following:

A sustained, luminous, and energetically resonant and coherent vibrational signal — which is a conscious, light-filled, vibrational signal that is sent forth with elevated thoughts and emotions in order that;

A coherent heart-brain entrainment.

The idea that you attract from the unified field of consciousness the highest possibility available in that field.

The idea that the energy of your vibrational signal, which <u>qualifies</u> the return of the gift from the unified field of possibility, must be at the highest level of vibration.

Charles Fillmore says that when you are residing in this level of vibration and awareness, you are working with and as the law of justice and the law of infinite expression or abundance. In the state of infinite expression of abundance, the awareness or consciousness from which you [share] is that which qualifies its own return.

Thus, it is the state of awareness from which the sharing is done in consciousness that creates the quality and level of luminosity and dynamic energy of expression. Then that luminous energy goes forth to attract a vibrational match in the unified field of consciousness.

Fillmore called this unified field, the kingdom of heavens — note the plural. What you send out as a vibrational broadcast signal of thought/feeling will attract back to you in accord with the level of luminosity that you sent it forth with and that will qualify its return to you. In addition, always let your return be qualified by the law of infinite expression, which calls forth the highest good for all — not just self.

You can create a higher vibrational luminous atmosphere by practicing inspired sharing. Inspired sharing is initiated through the simple act of self-observation. You do this by intentionally and honestly observing the thoughts and feelings that you are experiencing as you; share your gifts, make a purchase, or pay bills. After engaging in honest self-observation, engage in honest inquiry. Do this inquiry by asking yourself, "Am I sharing this gift, or doing this action, from a sense of joy and a sense of knowing that I am a luminous expression of abundance?" Wait to hear an answer, if it is yes, continue on! If not, stop and readjust your thinking and feeling!

2. What does Malachi 3:10 metaphysically mean in relationship to a consciousness of abundance and being an arc of infinite abundance in expression?

    Malachi 3:10, when understood from a higher dimensional metaphysical perspective, ignites a spark of luminous awareness that is a catalyst for the transformation of consciousness, through the shattering of old beliefs and perceptions.

> *"Bring your full tithes into the warehouse.*
> *So that I might have meat in my house.*
> *Prove me now says the Lord of Hosts...*
> *And see if I will not throw open the windows of heaven*
> *and call forth an abundance of blessings."*
> Malachi 3:10

Bring your full tithes into the warehouse

What is the *warehouse*? Metaphysically, it is your consciousness! And you are being asked to bring the *full tithe* into it. That begs the question then: What is a full tithe, when a tithe, historically, is 10% of 100% of something?

Metaphysically interpreted, when you bring your full tithe to consciousness, that tithe is not money or goods but a commitment to share a full 10% portion—of your thoughts, feelings, perceptions, and beliefs in higher awareness—towards a positive-orientation.

Source is requesting that you hold consciously, for just 10% of your day, higher positive thoughts, feelings, and spiritual principles in mind. It is a tacit agreement to unwaveringly and steadfastly holding higher positive-oriented thoughts and feelings, and to be nonresistant, not attached to outcome, and to not judge by appearance. In this state of awareness, you share your gifts, not from a need or a desire to get for having given, but from an established principle of abundance in consciousness, an awareness that has no doubts that—you cannot out give God.

So that I might have meat in My house.

So that I (I AM—not you, at the third-dimensional level of personality—but, higher dimensional awareness—True Self), may have *meat* (nourishing energetic substance, divine source energy derived from higher dimensional thoughts, feelings, emotions, and awareness which are then available to do good works in the world through a consciousness of abundance) in *My* (my-belonging to Source/God, not personal or individual) *house* (consciousness—a consciousness steeped in higher awareness that permeates the cells and cellular memories of the body temple).

Prove me now says the Lord of Hosts…

Prove me (trust unwaveringly, allow Me/Inner Source/Spirit to reveal and attest to what a small amount (10%) of dedicated inner work it takes to begin to express, share, and be that living consciousness of demonstrated abundance in all you do), *now* (do it now, in this moment, and not when you have more, know more, think you are more), *says the Lord of Hosts* (your inner intuitive guide/I AM that ever speaks to you as you are willing to listen to Its counsel).

And see if I will not throw open the windows of heaven and call forth an abundance of blessings.

Throw open the windows of heaven (vibrationally connect in accord with your vibrational signal to the highest possibilities available in the unified field of consciousness). And *call forth an abundance of blessings* (inspired Spirit-sharing calls forth infinite possibilities of greater good from the unified field of consciousness, more than you can ever imagine)!

Right use or Reverse use — the Law works!

~

I am a luminous arc of infinite abundance
expressing in the world!

A&A Inspired Sharing: A Spiritual Practice

Practical application of spiritual principles, through a conscious, intentional, and repeated spiritual practice, supports you in being a conscious, luminous, vibrational attractor of higher possibility; an inspired sharing spiritual practice is a catalyst for that to happen. It also, has the capacity to support you in the generation of a new vibration; a vibration that will align you with the highest possibilities available in unified field of consciousness. Inspired sharing is intentionally and vibrationally ignited when you engage the following:

1. First and foremost, acknowledge, know, and claim; *I am a vibrational being, living in a vibrational universe.*

2. Prior to the sharing of your first inspired gift, consciously remember that you are an arc of infinite abundance in expression. To support you in this, create a personal intention statement, in the form of an abundance statement. Consciously connect with your statement, several times a day.

   I am infinite abundance in action, and the universe attracts to me in alignment with the vibration of my claim! NOW!

3. Select a spiritual place (a ministry, temple, minister, or spiritual leader) who is supporting your soul growth and evolution. Have clarity when making your selection. If you are unsure of, or do not know where to share, please consider sharing your gifts, or a gift, with the author's ministry.

   Awakening Awareness Ministry, c/o Rev. Toni G. Boehm
   430 N. Winnebago Dr.   Greenwood, MO 64034

Prior to each inspired sharing: consciously pause; breathe; go to heart and connect with the resonant field of energy there within; affirm your connection, I am a luminous point of energy, consciously creating an arc of infinite abundance in expression.

Remember, it is the level of your vibrational energy, in that moment of your sharing, that qualifies the return of the gift to you, from the unified field of possibility.

7

Soul Mastery in Action

I Am That I Am.
Divine and Eternal Life is built into the
body of the manifest human and through thinking
(regeneration through the renewing of the mind)
we can call it forth as a "living" reality.
Let those with ears, hear!
Charles Fillmore

If one where to judge by appearances, with all fighting and upset present on the world scene, it would appear that humankind as a collective consciousness is not vibrationally ripe, nor emotionally ready, to fully align with the message of higher dimensional awareness, or its frequencies. At least not yet, but have faith, it is coming!

As each individual consciously chooses to ascend the spiral of conscious evolution and adds that ascended consciousness to the collective, the soul mastery frequency will become the dominant energy of the collective mind of humankind and all of life—it cannot be any other way, that is how evolution works. In this higher dimensional awareness and frequency, people will live in abundance and share freely, for they will know how the principle of abundance works. They will have the capacity to do what the master teacher Jesus did with the loaves and fishes. That is the impact that the vibrational signature of 5th dimensional frequency and higher as a conscious state of being holds.

The shift for humankind, as a collective consciousness, starts with you! When you make the conscious choice to apply the principles that support the development of soul mastery; when you consistently hold them as a way of being; then the energy, vibration, and frequency that supports soul mastery will embody itself in you; and you will begin to notice that you are a master facilitator of energy! And as you transform, you transform the world! Enjoy the journey of soul mastery!

The wisdom-teachings, energy, initiations, and experiences associated with Soul Mastery are initiated by your yes, your agreement, to participate consciously with them. As you say yes, you are calling forth and initiating the occurrence of many of following elements, initiations, transformations, and/or ideas:

- A personally designed soul evolution journey and process that will lead to a release, and subsequent conscious transformation of the dense energy residing in your energy field;
- A greater understanding of how energy, vibration, and frequency work in your life;
- The sacred union of divine love (feminine) and divine wisdom (masculine), and the accepting of all the experiences that are required to bring that to fruition. This sacred union, or internal balance of your masculine and feminine natures, births higher dimensional awareness and conscious neutrality, along with the capacity to live from a state of detached compassion, balance, harmonic resonance, authenticity, transparency, and more;
- The establishment of conscious neutrality, as an established state of being and awareness is important for it has the capacity to shift the energy, vibration, and frequency within your physical, mental, emotional, spiritual, and auric fields. This transformation results in having the capacity to stay present in the present moment, to be the energetic presence of all is well, and to alchemically transform energy fields, through your presence when conditions are right;
- Being a conscious demonstration of the spiritual principle of abundance; a repeatable and sustainable demonstration;
- Knowing, without a shred of doubt, that you are the vessel through which higher dimensional awareness is being shared in the world.
- Knowing that you are a master facilitator of energy, and soul mastery in action, and that all your actions arise from and are inspired by, higher dimensional awareness, Source Energy.

At a level of structure, all is energy and frequency, and it is an energetic frequency that is identical to Source-Creator-God.
You were born from and created in this frequency.
The physical form you stand in was created by you in order that you might partake in experiences here on this plane of existence, and that you might ascend in consciousness through the lessons gleaned from those experiences. It is the physical form that is the vehicle through which you, and all, will ascend — appreciate and honor it! [41]

---

[41]A compilation by Toni G. Boehm, and other master teachers

8

Alignment and Attunements Experiences:

A&A: Social Action as Viewed Through Higher Dimensional Awareness and Soul Mastery

You can't see a new reality, or move into higher dimensional reality, until you do the work to rent the illusionary veils that holds the information of that new reality at bay. With this is mind, consider the possibility that the following scenario is a metaphysical treatise on the next phase of evolution regarding spiritual, social action. It is a how-to on energetically participating in spiritual, social action from a soul mastery awareness.

As you engage in this exercise, you are invited to put all pre-prescribed ideas and notions you hold about the man named Jesus aside. It doesn't really matter whether he lived or if he did or did not do all the things "they" say he did! What is important is that through the legacy of his story and life experiences—even if they are a myth—are woven mystical teachings that are discoverable by those who have *eyes to see.*

John 8: 1-11 Jesus and the Woman Caught in Adultery [42]

1 Jesus returned to the Mount of Olives, **2** but early the next morning he was back again at the Temple. A crowd soon gathered, and he sat down and taught them. **3** As he was speaking, the teachers of religious law and the Pharisees brought a woman who had been caught in the act of adultery. They put her in front of the crowd.

4 *"Teacher,"* they said to Jesus, *"this woman was caught in the act of adultery. 5 The law of Moses says to stone her. What do you say?"*

6 They were trying to trap him into saying something they could use against him, but Jesus stooped down and wrote in the dust with his finger.

7 They kept demanding an answer, so he stood up again and said, *"All right, but let the one who has never sinned throw the first stone!"*

8 Then he stooped down again and wrote in the dust.

9 When the accusers heard this, they slipped away one by one, beginning with the oldest, until only Jesus was left in the middle of the crowd with the woman. **10** Then Jesus stood up again and said to the woman, *"Where are your accusers? Didn't even one of them condemn you?"* **11** *"No, Lord,"* she said. And Jesus said, *"Neither do I. Go and sin no more."*

---

[42] *Holy Bible, New Living Translation, Christian Testament*

With Jn. 8: 1-11 as the foundational context for your responses regarding spiritual, social action, ponder the following ideas:

- Consider, was the man named Jesus, in this scenario taking a stand for spiritual, social justice and action? If your answer is yes, ask yourself:
  - What influenced the outcome? Was Jesus taking a stance that was influenced through resistance, or a connection to the flow of inspired action—in alignment with higher dimensional awareness?
- Consider what he was doing, energetically, when he stooped down to write in the sand with his finger.
  - Was he perhaps appearing to be passive as he aligned with Spirit?
  - Was he perhaps, consciously grounding negative energy?
  - Was he taking action from his adverse ego, or spiritual ego/I Am?
- Consider what the scenario means as a context for alchemical transformation.
  - As a conscious, energetic field of nonresistance and non-defensiveness, might have Jesus alchemically transformed the energies in the field around him? Might he have, through an invisible "spark" of interconnection, shifted the mind-fields of those around him, and released the anger?
- Consider when he said, "Let he who is without sin, cast the first stone." "Where are your accusers?" "Go and sin no more."
  - Was what was said important or, was it the consciousness from which he said it?
  - Why is it important that old ways, habits, and beliefs be left behind?
- Consider this the next time you want to take a spiritual, social action.
  - What is the consciousness from which you want to take action?
  - "Fight-thinking" or "Right-thinking"?

To ascend to the level of the conscious, compassionate engagement that the master teacher Jesus revealed that conscious action must be taken without anger, pity, seeing a need to fix, or trying to save someone. Also, that holding a level of vibration that can alchemically transform a situation requires the development of spiritual mind discipline, an overcoming of the adverse ego, and the development of conscious neutrality, as a natural state of being.

Are you ready to be a master facilitator of energy? Are you ready to live from the higher dimensional awareness of soul mastery? If your answer is yes to these questions, know you have the tools inherent within you, draw them out. *Go and "sin" no more* — go forth and do not forget the Truth of your being! I am conscious neutrality in action! I am an alchemical transformer of energy! I am a master facilitator of energy!

A&A: Karmic Density Release [43] *– Meditation A*

Breathe and center yourself in your heart space. Imagine your physical body is surrounded by a luminous light body. Sense it, see it, feel it! Now take a breath and project that light body out in front of you. See your light body clearly, standing in front of you.

Imagine that all the experiences, interactions, and exchanges that have occurred in your life are appearing before you and, as they do, each morph into a grain of sand. Before you are grains of sand measuring into the thousands. Imagine that each grain of sand begins to fall through the spaces of the rays of your light body. Allow the grains of sand to fall toward your feet. How much of your body do these grains of sand fill? One-third, half, three-quarters?

Once the sand is settled, notice that a sticky, mucky, black liquid begins to appear between the grains of sand. This sticky muck touches each grain of sand and it congeals the sand into clumps of a dense and thick substance. This substance represents the muck and karmic density that is formed in you as you cling to the emotional residue of past third-dimensional experiences. This muck and karmic density remain until you choose to release it.

This is the choice point moment! See yourself taking a big, deep, expansive, magical, and mystical breath. A breath that has the power to blow away all the heavy substance holding the grains of life experiences in place.

Now blow that breath out and, as you do, watch the muck dissolve through the power and essence of life held within your breath. All that is left is the grains of sands, floating free in your light body.

---

[43] *Adapted from M. Sheridan. YouTube. Awakened and Empowered.*

Now that the muck and karmic density residue of the past is cleared, you can use the grains of your life's experiences in a new way. You can utilize them as energetic reference points to make new choices and to expand your awareness. You can look at your experiences and say, "Ahh, yes, I have had this experience. I know what happens. I know the muck and karmic density it can accumulate. Therefore, I consciously choose to make a new choice and to create a new result."

Each choice you make toward conscious shift energetically does the following: it frees up muck and karmic density residue; it lifts and expands your level of vibration; and it sets more intercellular light free, which brings in new information regarding who you are!

A&A: Karmic Density- Primordial Meditation B

Releasing the muck of perceived attachments and karmic density frees up energy and supports you in perceiving reality from a new perspective—and that raises your vibration.

Take a deep breath and bring your awareness into your heart space. Relax into this space of spirit and connection to the unified field of consciousness. It is the place where you claim and align to the truth of your being. In a state of connected awareness, you are receptive to higher dimensions of awareness and understandings of truth.

In this state of interconnection and being, consider a person with whom you have or have had issues, sense you are karmically tied to, or someone you are ready to be at peace with.

Once the name or perception arises in your imagination, imagine them dissolving into their energetic pre-birth primordial vibration. Imagine them as if they had never been born, never had come into form, never been invested in a physical self: just a vibration. As you imagine them in this primordial vibration state, ask yourself:

- If they have no body, are they still the person I know?
- Without a body, are they still who I believed them to be?
- Have I mistakenly perceived them to resemble what I think they are?
- Have I created a perceptual form that I now hold on to?
- What changes as I see them as a vibration at the level of pre-structure, a place where we all are pure truth?

Standing as sacred witness to this being's primordial vibration, you are invited to make the following claim. Repeat it three times:

I stand as the holder of higher vibrational awareness, the place of alchemical transformation, a place where energy shifts and transforms lives instantly.

In this energy field, I know who you are, and I now agree to see you in your divinity. I see this in place of other things or experiences we have had together.
I stand as the witness to this in Truth and to this Truth.

Breathe. And when you are ready, return to the present moment.
Take a moment, then ask yourself:
What do I now notice, energetically, about me? The other person?
What has shifted in my perceptions regarding them?

A & A Micro-Meditation—An Embodiment Practice

Micro-meditation is a spiritual practice, that once developed, provides an instant support for you on your journey! A spiritual practice does not have to be an hour long to be effective. It is more about how you intentionally focus your attention and energy towards the practice, than the amount of time.

In this, micro-meditation embodiment practice, you are intentionally focusing your attention and energy in the heart space, and then, on connecting with the energy that is "alive" and resides in the heart space. Energies such as, the spaciousness of emptiness, the Unified Field of Consciousness, and the crystalline Christ Energy (crystalline/very clear).

This is both a practice for the development of soul mastery, and conscious neutrality. Remember, when you open yourself to the higher dimensional awareness that holds the state of being conscious neutrality in action. Remember, neutrality does not mean that you do not care, it is about—not carrying any longer—lower vibrational energy, or negative emotional residue. When you consciously connect with the Christ/Source Energy/the Field for even a brief moment, you become one with that Field, you hold that energy, and people notice and experience that energy emanating from you. And when you are this energy in action, this is the greatest act of sacred service you can perform. For you are an energy changer, in service to the greater good of all. It can be no other way, for this is what this Field of energy holds.

When you consciously surrender, over and over, into the field of higher dimensional awareness, Spirit/Source begins to guide your actions, creates inspiration and inspired thoughts and actions, and these will all arise from a level beyond, reason and intellect, even illumined intellect.

<u>Instructions</u>:

First, create an intention. Intend to commit to this spiritual practice, until you know that you are consciously in alignment with the Field/Christ Consciousness. Hold also, a commitment to engage with the spiritual practice at least twice per day, or more, for a minimum of 7 days. (Remember, this is not about a long practice, it is the development of an intuitive sensing and embodiment practice.)

The first several times, as you engage the practice, it may take five or ten minutes, or longer. Then as you consciously connect with the Light Energy, over and over, it will become an automatic response mechanism that you consciously tap into, with your eyes open, in the moment of a heartbeat. Personally, I find that is akin to the ancient *Prayer of the Heart*, practice. Where the pilgrim, seeking to know how to pray without ceasing, asked the wise one, "How often do I have to do this prayer? And the wise one replied, "A thousand times a day, until the Prayer, begins to pray you."

The Micro-Meditation Embodiment Practice

Close your eyes.

Take a deep breath and follow the energy of the breath into the heart space.

Imagine that as your breath reaches the heart, it lands in the midst of a ball of golden light.

It is the field of crystalline Christ/Source energy/unified field of higher awareness, that resides within you, at all times.

Your breath is moving in and out, in the midst of this field of golden energy. You are one with this field of golden light, this crystalline Christ energy.

Consciously, expand the light.

As the light expands, so does your essence, along with it and you are aware that it is filling your entire heart space.

Now it is expanding out further and filling your entire being.

The expansion continues, beyond the artificial boundary of body — it is expanding it is affecting your cellular structure, and your DNA — light is expanding out from them.

Gently, with every breath, imagine that you are pushing the light out into the room.
Fill the room.
Then reach beyond the room.
Finally, reach out into infinity.
Expand, expand, expand the light.
Hold in the Silence!
When you are ready, take a deep breath, and gently, open your eyes. Know the you carry the energy of this experience with you, now and forever.

A&A: Alchemical Transformation Witnessing of Another

How do you begin to practically apply these ideas and constructs in your life? One simple way is the process known as alchemical transformation witnessing. Alchemical transformation witnessing is the conscious witnessing of another from a higher level of awareness instead of witnessing who they appear to be.

How does alchemical transformation witnessing work or happen? Let's start with how it doesn't happen. It doesn't happen when you see or perceive an individual (or a group) to be who and what you think they have always been: she is bossy, he is overbearing, she is so weak, she is controlling, he is greedy, or they are evil. When you do that, all you are doing is reaffirming what you have thought or believed to be true in your thinking as a past perception about them.

Knowing and thinking are the not the same thing. When you *know* someone as truth, you are investing in thoughts that arise from a higher dimension of awareness—the level of truth.

When you think about someone from your past perceptions, you are investing in your own perception of what you think they are, which ultimately means nothing. We are all the same at the level of structure.

When you look at someone and witness them as a mother, father, brother, sister, enemy, rival, the one who did this or that, you are witnessing them based on a past perception of knowing in time, based on form in personality.

When you make a claim for another   I behold the Christ in you! I behold the Buddha nature in you! I behold the High Truth in you! You are making the claim that you know who they are in higher dimensional awareness, in truth. I see you outside your physical form. I see you at the primordial level of structure. I see you as pure being.

All this simply means that you witness them outside of a pre-prescribed structure of perception that you have assigned to them based on previous experiences and even previous learnings given to you by others. Predefined structures of knowing—such as the "always late one," the "angry one," the "bossy one," the "one who can't be counted on," the "immature one," the "uncaring one," the "selfish one," and so on—color your view of truth and reality.

In a moment of conscious allowing, of being receptive to a higher truth, you align them to what they truly are. As you witness them differently in consciousness, you begin to see them differently.

Understand the difference here! *I behold the Christ in you* doesn't mean I think I know this, or I think I know who you are. It is not a platitude you created.

Until you understand the magnitude of this and have a clear realization of the deep meaning of, I behold the Christ in you, and I know who you are in Truth, you are just playing a game of platitudes. Who you think anyone is, is always a false identification based on perception. This is not about looking at a person who is not well and seeing them well because that is what you wish for them. No! This is bigger than that!

This works whether it is about witnessing a person differently, seeing wholeness, seeing abundance, or any other spiritual principle as a *living* truth. This is a high-level awareness of the capacity to recognize the divine and inherent worth as the divine. By claiming and holding that awareness of truth, an alchemical transformation takes place in the energy field.

Note: a learning (and unlearning) guide for a year-long curriculum is currently being prepared. Please email, btoni33@yahoo.com, or visit www.toniboehm.com for more information.

About the Author

Toni G. Boehm, M.S.N., Ph.D., P.C.C.

Founder: Academy of Leadership Excellence
Facilitating Alchemical Transformation Through
High-Performance Coaching, Leadership Team Development, and
Spiritual Direction

As an internationally known inspirational and dynamic speaker, author, certified organization and life coach, nurse, and poet, Boehm reveals an expertise in coaching skills, group facilitation, and an ability to create transformational experiences. Boehm and her husband have five children and six grandchildren, and one absolutely gorgeous, great-grandchild.

Experience:
- Organizational Coach for Leadership Skills, & Conflict Resolution Consultant
- National Ministry Leadership Skills & Conflict Resolution Team Coordinator
- Founder & CEO, Academy of Leadership Excellence
- Professional certified coach, organization and individual (ICF)
- Academy of Coaching Excellence, Licensed Academy Trainer
- Keynote speaker at multiple national conferences
- Vice President Education, Hospitality, & Retreats – Unity Institute
- Dean of Administration, Unity School for Religious Studies
- Coordinator for annual American Nurses Association national conference
- Sr. Staff Specialist for best practice research & standards for Credentialing and Accreditation for nursing (American Nurses Association) and for ministry leadership skill development and support (Unity Worldwide Ministries)
- Hallmark Cards Inc., Nurse Practitioner
- Interim Director of Nursing and Nurse Practitioner, Truman Hospital

Awards and Achievements:
- University of Missouri Woman's Council Award for Outstanding Master's Thesis Research. Research on identifying stress in executives was selected for presentation at John Hopkins Hospital Annual Medical Symposium.
- *Who's Who in American Women* and *Who's Who in America*
- Elected for two terms to Missouri State Board of Nursing
- Published author of twelve books, multiple CDs, and numerous articles
- 2014 Volunteer of the Year Award, Unity Worldwide Ministries
- 2017 Recipient of the Charles Fillmore Award for Visionary Leadership

"Coaching to possibility and the yes' of life!"

CPSIA information can be obtained
at www.ICGtesting.com
Printed in the USA
FSHW020059221019
63245FS

9 781645 163251